*An Introduction to the Thoroughbred*

PETER WILLETT

# *An Introduction to the Thoroughbred*

Foreword by

THE 16th DUKE OF NORFOLK,
K.G., P.C., G.C.V.O.

STANLEY PAUL, LONDON

STANLEY PAUL & CO LTD
3 Fitzroy Square, London W1P 6JD

An imprint of the Hutchinson Publishing Group

London Melbourne Sydney Auckland
Wellington Johannesburg Cape Town
and agencies throughout the world

First published 1966
Second impression 1969
Revised edition 1975


Set in Monotype Times

Printed in Great Britain by The Anchor Press Ltd
and bound by Wm Brendon & Son Ltd
both of Tiptree, Essex

ISBN 0 09 123740 8

# *Contents*

# *Illustrations*

# *Acknowledgements*

It would be impossible to mention all the people who have helped, by the written or the spoken word, to germinate the ideas which find expression in this book. However, I must not fail to acknowledge with gratitude the assistance of two men, Mr. E. J. C. Blackwood and Mr. Patrick Saward, whose researches have illuminated specialised areas of thoroughbred history and who have kindly placed unpublished material at my disposal. Mr. Blackwood's compilation of winning sires statistics from 1721 onwards has facilitated the task of evaluating the contribution of individual stallions to thoroughbred evolution, particularly in the period up to 1850; and the work of Mr. Saward has enabled me in the present volume to amplify my earlier account of the inbreeding experiments conducted by the 12th Earl of Derby in the first quarter of the nineteenth century.

# *Foreword*

BY THE 16th DUKE OF NORFOLK, K.G., P.C., G.C.V.O.

In January 1965 I invited Mr. Peter Willett to join Mr. Geoffrey Freer and myself to form a Committee set up by the Jockey Club to report on the 'pattern of racing'. We worked together from March till the end of June to produce a report which has received much publicity. I count it a great privilege that so soon after this event the author should ask me to write a foreword for his new book *An Introduction to the Thoroughbred.*

As the future pattern of our racing has been the subject of so much discussion, the arrival of this book adds a great deal to the general interest of our breed from which the racehorse itself is derived.

At one stage of his work Mr. Willett has analysed at some length the many and varied theories which have been produced through the years by those who have given much time to answer the question: How to breed a good racehorse.

In later chapters readers will find his views concerning the French and American methods of breeding and racing. In fact every aspect of this intriguing pastime is fully covered in a most interesting manner.

The amount of his own knowledge is considerable and the vast research into the past with all its problems makes one realise the work he has undertaken.

As one comes to the end one may well ask oneself: 'Why do we go on trying?' Because the answer still is not there as to how we can make sure of breeding a first-class racehorse. But if the answer were to be there then all breeding and racing would be dull. It is the uncertainty of both that urges one on and the excitement when something better turns up.

Many there are who would have done my job better, but it is in the reading of the book that the pleasure is to be found. Mr. Willett deserves our thanks for making a valuable contribution to the literature of the Turf.

# *Introduction*

It seems best to begin by setting down some of the things that this book is not intended to be. This book is not a manual of stud management; it is not a work of reference containing details of the careers of the principal stallions and mares who have influenced the development of the British Thoroughbred; and it is not a vehicle for further 'infallible' systems for breeding more and better winners.

The aim rather is to provide a consecutive account of the development of the British Thoroughbred and of the principles involved, from his known origins down to the present day. For this purpose I have endeavoured to avoid breaking up the narrative with those formidable accumulations of statistical matter from which the reader is liable to shy away like an unschooled hunter from a five-barred gate. On the other hand it is impossible to give even the most superficial account of the evolution of the British Thoroughbred without attempting some elucidation of the principles of heredity, and no apology is offered for the portions of the book which deal with this aspect of the matter.

Wherever possible the argument has been advanced by drawing upon examples rather than straightforward exposition. The examples chosen may not be the only or the best ones available, but are those which appeared to fit most readily into the plan of the book.

There is a tendency, even among those who are well informed about the development of the throughbred in modern times, to disregard—indeed to treat as unworthy of notice—that initial period of thoroughbred evolution which may be described as 'Old Bald Peg and All That'. This attitude, if maintained too rigidly, may be mistaken, because the development of the thoroughbred from the earliest down to the present time has been a continuous process depending on the same principles of heredity, and the study of the way that limited resources were used to establish the breed may provide the clearest picture of the principles in action. In the same way we may be able to learn more about the basic techniques of garment making by watching a woman knitting a pullover than by inspecting a fully mechanised and automated clothing factory.

Breeding racehorses is, and has been for a long time, an expensive business or hobby, but the rewards for success at the highest level are great. It has been described fairly as an inexact science; in every mating involving the thoroughbred there are imponderable factors which are apt to upset the most brilliantly conceived plans. Not surprisingly the inexact science, with its financial overtones, has been encrusted with a vast lore of catch-phrases, hare-brained theories and miscellaneous short-cuts to success; and much of the lore is built on the flimsiest, or non-existent, scientific foundations. Many breeders of the present time are either sceptical of the lore or reject it altogether, but others are reluctant to abandon theories which have been more or less sanctified by tradition and endless repetition. As late as 1964 one distinguished breeder was writing of the impact of the teaching of geneticists on the cherished theory of the perpetuation of family characteristics: '. . . after all, what advice of really practical value have these experts been able to give to the average breeder?' This rhetorical question seems to be inviting the answer that fiction is preferable to fact, if fact involves the destruction of comfortable beliefs. The proper answer should surely be that the primary function of geneticists in respect of the thoroughbred must be to provide scientific standards by which every breeding theory, however plausible, should be judged.

The reasons why people begin to take an interest in racing and breeding are many and various. My own interest originated when I noticed the name Knight of Knockeevan, a marvel of mellifluous alliteration, in the racing column of a newspaper. Knight of Knockeevan was a good jumper, but he was also a gelding, so he inspired no subsequent interest in breeding. The next year Hyperion won the Derby and, as his owner Lord Derby was chairman of the governors of the school I then attended, we were given an extra week's summer holiday to celebrate. After that there could be no looking back, because Hyperion became one of the great sires of the century.

The gambling instinct is, of course, the main reason why most people take an interest. Yet the gambling instinct alone cannot account for the extraordinarily widespread appeal of horse-racing since horse-racing is an inefficient medium for gambling in some respects. It provides less frequent opportunities for turning money over than cards, or roulette, or greyhound-racing even; it provides less glowing prospects of winning a fortune for an outlay of a few pence than the football pools. Instead it provides that many-sided fascination which is summed up in the phrase: 'The glamour of the Turf'. And not the least absorbing aspect of that glamour is the continuity of interest begotten by the transmission of the great qualities of famous parents to their offspring. When a Derby winner, Owen Tudor, is found to have another Derby winner, Hyperion, as his sire, a resonant chord of memory is struck in the minds of most followers of racing.

Even on the level of winner-finding pure and simple, the study of breeding has a part to play. This relevance may be conveyed by the definition of heredity as 'the tendency of like to beget like'—or, put the other way about, of offspring to resemble their parents. If we know something of the characteristics of the parents, grandparents and other close relatives of any given horse, we may be able to make useful deductions about the characteristics likely to be possessed by that horse. These deductions may cover aspects of the make-up of a racehorse like speed, stamina, ability to act on various types of going, temperament, constitution and soundness. The breeding of the animals engaged may never be the principal or decisive factor in assessing the chances in a race, but it is always one of the factors that ought to be considered.

The history of the Derby provides the most striking example of the relevance of breeding to finding winners. Of the thirty winners of the Derby from the end of the second world war in Europe down to 1974 only four – Hard Ridden, Relko, Sir Ivor and Roberto – did not have animals capable of high class form over at least 1¼ miles as both sire and maternal grandsire.

A large proportion of the runners in the Derby may pass this test, but at least the application of it would induce a sceptical attitude to the claims of a horse like Ki Ming, who started favourite for the 1951 Derby after winning the 2,000 Guineas. Ki Ming was by the pure sprinter Ballyogan, and was beaten at about halfway at Epsom.

Some elementary knowledge of breeding should thus have a place in the mental knapsack of every follower of racing. That is a strictly utilitarian view of the subject. Apart from its value as a complement of the form book, the study of how and why the thoroughbred has evolved may be regarded as intrinsically satisfying and worth while for its own sake. It is in this latter spirit that the following pages are offered to the reader.

# 1

# *The Origins and Development of the Thoroughbred*

THE ORIGINS OF THE BRITISH THOROUGHBRED, who became the principal and practically the sole source of horses acknowledged as 'thoroughbred' in all countries in which the sport of horse-racing flourishes in the modern world, are shrouded in mystery. We have to make do with the few facts that have been established, but we can be sure that there was a period of intensive evolution, lasting roughly from the middle of the seventeenth to the first quarter of the nineteenth century, which resulted in the emergence of the thoroughbred as he is known today. In the course of this evolutionary process the average height of the racehorse increased by about a hand and a half, or six inches, and there were corresponding increases in overall size and physical scope, strength, length of stride and speed.

Acceptance of the fact of this remarkable development of a specialised breed of racehorses is bound to pose the questions why and how it came about. As to the question why, clearly there must have been some powerful influence to initiate the process, and it may not be necessary to look beyond the Stuart Restoration of 1660 and the personality of King Charles II to find it. Charles was the first royal racing regular, and was a competent race-rider himself. His unwavering enthusiasm and constant patronage of Newmarket helped to make racing a national sport. This is, of course, not to say that horse-racing was an innovation in Restoration England. It had existed for centuries, but as a purely local sport enjoyed in dozens of places up and down the country, and without any central organisation. Nor did Newmarket become famous as a sporting centre for the first time after 1660. The little Suffolk town of about two hundred houses, tucked away in a fold of the broad, undulating heath, had been frequented by both the first two Stuart kings, James I and Charles I, but they had used it primarily as a hunting lodge. It was the conjunction of Newmarket, horse-racing and the character of Charles II that inaugurated a new era for the sport.

Charles used to spend weeks at a time at Newmarket. He had the ruined palace there rebuilt by Wren, and on his visits was usually accompanied by a large part of the court, a contingent of his ministers and an assortment of mistresses, notably Nell Gwynne and Louise de Kerouaille. Lord Arlington, one of the ministers of the Cabal, used to entertain lavishly at nearby Euston. At the same time racing was beginning to develop as an organised sport. The courses on Newmarket heath were marked out by lines of white posts, and the King assisted in framing the conditions of races besides acting in a judicial capacity in racing disputes. Betting was heavy and competition tremendously keen.

Inevitably this close connection between racing and the sources of social influence and political power conferred an immense new prestige on the sport. Moreover, Newmarket, apart from the unique excellence of its courses and training grounds, had a strategic position sixty miles out of London in the direction of Yorkshire, the principal horse-breeding county of England. Nothing could have been more conducive to the improvement of racehorses than this accessibility of Newmarket to both London, the permanent home of the court, and Yorkshire, for in this way the growing spirit of competition was communicated to the breeders of the north. What Charles II had begun, Tregonwell Frampton, who held the post of trainer-manager to William III, Queen Anne and the first two Georges, continued. Frampton, a skilful race-rider like Charles II, used his experience and his royal appointment to maintain and advance the prestige of Newmarket as a racing centre and the focus of north–south competition, and so became the indispensable link between the post-Restoration period and the second half of the eighteenth century, when the Jockey Club began to emerge as the dominant force in the world of horse-racing.

It was through this fortunate chain of circumstances that the incentive to breed improved racehorses, which had been provided by Charles II's love of racing and association with Newmarket, was sustained during the whole of the formative period of the thoroughbred. At the end of the period the thoroughbred was fully established as a breed that was not only distinct but incapable of further improvement by crossing with other breeds, and the writ of the Jockey Club was beginning to run throughout English racing. There was no coincidence about this parallel development. The two things, the breed of horses and the centralised control, were the two sides of the same coin which had been struck in Restoration days at Newmarket when Charles II gave judgement in a racing case between a courtier and a Yorkshireman.

In laying the foundations of racing as a national sport Charles II may have made his most lasting contribution to the English way of life. It may not be extravagant to claim that without him there would have been no thoroughbred, no Jockey Club, no General Stud Book, no Racing Calendar, no Classic races, no betting shops—none of the ingredients of the system of horse-racing as we know it.

Charles II, then, gave the incentive to improve racehorses. How that improvement was brought about is quite another matter, and it must be obvious that success could never have been achieved if ideas about livestock improvement had not been current among the breeders and landowners of the time. The fact is that the eighteenth century, within which the most dramatic progress in the evolution of the thoroughbred was made, was an age of agricultural revolution in England. In the course of the century the population doubled, but English farmers raised their efficiency to such a remarkable extent that they were still able to feed it. The chief weapons in the drive to increase production were improved methods of cultivation, the introduction of new crops and, most important for the thoroughbred, the beginnings of scientific stock-breeding. Men like Robert Bakewell of Dishley came to the fore, propagating the belief that livestock could be improved and brought ever nearer to an ideal type by selection and inbreeding. Bakewell put his ideas into practice with outstanding success, producing the Leicestershire breed of sheep, the Leicestershire breed of long-horn cattle and, drawing closer to the thoroughbred, a breed of black cart-horses remarkable for their strength and endurance. He showed the way ahead to breeders with equal clarity when he founded a club, the Dishley Society, for the express purpose of ensuring purity of breed.

Bakewell also studied the problem of the relation of food intake to improvement in meat cattle, and his example led to a greater awareness of the importance of correct feeding of all livestock. As a result, doctrine concerning the proper diet for horses in training was radically altered. Gervase Markham, writing before the agricultural revolution, gave the following recipe for a kind of bread on which a horse should be fed during the final stages of his preparation for a race: 'Take a strike of beans, two pecks of wheat, and one peck of rye, grind these together, sift them and knead them with water and bran, and so bake them thoroughly in great loaves, as a peck in a loaf; and after they are a day old at the least your horse may feed on them, but not before.' William Youatt, writing about the time when the evolution of the thoroughbred was nearly complete, had very different ideas on the subject. He recommended oats as the basic ingredient of the horse's diet, and insisted that they should be old, heavy, dry and sweet. He also investigated the properties of other cereals as food for the horse, pointing out that barley and wheat were both more nutritious than oats, but concluding that there seemed to be 'something necessary besides a great proportion of nutritive matter in order to render any substance wholesome, strengthening, or fattening'. Horses fed on barley were liable to inflammatory troubles, and horses fed on wheat to indigestion and obstructions of the bowels.

In this way scientific principles were applied to methods of feeding as well as breeding. Thus the evolution of the thoroughbred may be regarded not as an isolated, inexplicable phenomenon, but simply as one aspect of the gigantic forward leap that changed the face of animal husbandry during the eighteenth century.

### *The Raw Materials*

What type or types of horses were used to produce the new breed, the British Thoroughbred? This question has given rise to much acrimonious and, it must be said, often puerile debate. The special pleaders have been busy in defence of the claims of Arabians or a native English breed of 'running horses' as the principal source of merit in the thoroughbred, according to where their own loyalties have lain. Each party has tried to annex all the doubtful arcas in the encestry of the thoroughbred, and has been guided by blind faith instead of a desire to discover the truth. Unfortunately the doubtful areas have been large, because the early methods of recording pedigrees were unsystematic, inconsistent and frequently inaccurate.

There was horse-racing in Britain for many centuries before the Restoration. The horses used for the purpose were probably small, light breeds like Galloways and Irish Hobbys, but these had been modified by numerous importations, some directly from the East, others from Italy and Spain where Arabians had been mixed with heavier northern breeds. Observation of the modern thoroughbred suggests that his ancestry contains some ingredient other than the pure Arabian; while some strains, like that traced through Djebel, his son My Babu, My Babu's son Milesian and the progeny of Milesian, have the characteristic Arab quality, others have a distinctly coarser stamp.

Whatever the sources of the original stock may have been, the evolution of the thoroughbred was a wholly British achievement. Patriotic feelings ought to be satisfied with that assurance. Many of the raw materials used in a variety of British-manufactured goods are imported; the products are not considered any the less British for that.

Arab stallions were considered the best for getting racehorses before the Restoration. Markham recommended them, and the first Duke of Newcastle, writing during the Commonwealth, suggested the use of Barbs instead merely on account of the exorbitant cost of well-bred Arab stallions. He stated that a bad Barb was better for siring racehorses than any English horse. Happily, the fourth Duke was not bound by the same motives of economy, as he had two Arabian stallions, called the Great Arabian and the Little Arabian, in his stud at Welbeck early in the eighteenth century. The golden age for the importation of Eastern stallions was from the middle of the seventeenth to the middle of the eighteenth century, when the incentive to improve racehorses was doing its work and before the Arab cross had ceased to be effective. The exact numbers of the imported stallions, and the antecedents of many of them, cannot be discovered. They were liable to be labelled as Arabians, Turks and Barbs indiscriminately, and to be called after whoever owned them at the moment. Thus Honywood's Arabian was identical with Sir J. Williams's Turk and Sir C. Turner's White Turk, Sir Charles Sedley's Arabian was the same horse as the Compton Barb, and the Godolphin Arabian, one of the most influential of the imported stallions, was sometimes referred to as a Barb. It is fascinating to speculate how many individual horses were to be found under the descriptions of Hutton's Barb (brought into England by Mr. Marshall), Hutton's Grey Barb (present from King William), Hutton's White Turk, Hutton's Grey Turk, Hutton's Bay Barb (sire of Black Chance) and Hutton's Bay Turk. By way of introducing a further complication, Hutton's Bay Barb was identified with the Mulso Bay Turk. The researches of Lady Wentworth led her to believe that the Alcock, Pelham, Ancaster, Honywood and Sir Robert Sutton's Arabians and the Akaster and Holdernesse Turks may all have been the same horse, and we have already seen that Honywood's Arabian was identical with two other horses whose names appear in the records. The confusion is extreme.

The imported stallions also included a few Persian horses and one Egyptian, Croft's Egyptian horse. The general principle was the horses were described according to their countries of origin, but this principle was not adhered to strictly. The fact that most Arabian horses brought to England were shipped from ports within the Turkish dominions caused mistakes, and in any case most of the light horses in Turkey were pure-bred Arabs. Some of the horses called Turks were spoils of war taken from the Turkish armies during the campaigns in Austria and Hungary. The Lister or Stradling Turk, who was a very successful sire, was captured at the siege of Buda and imported by the Duke of Berwick in the reign of James II. Barbs came from Barbary, the part of North Africa comprising modern Tunisia, Algeria and Morocco. Some of the Barbs may have been of pure Arab descent, but others were of mixed breed, since there had been considerable traffic in horses between North Africa and Europe during the Arab occupation of Spain.

The General Stud Book listed 103 stallions that were imported or of entirely foreign pedigree. By adding other stallions mentioned by William Pick* and found in other records, the total may be raised to about 160 after eliminating the most obvious duplications. The true number was probably much lower. Whichever count is taken—the Stud Book or the Stud Book plus Pick and others—about 50 per cent of the stallions seem to have been Arabians and about 25 per cent each Turks and Barbs.

Of all these sires three founded dynasties that have persisted down to the present day, though, as we shall see from a study of the pedigree of Eclipse, this is not to say that their influence

* *The Turf Register and Sportsman and Breeder's Stud Book,* 1803

was paramount in the making of the British Thoroughbred. They were the Darley Arabian, the Byerley Turk and the Godolphin Arabian. Of these three male lines, the Darley Arabian has far outstripped the other two in numbers during the last hundred and fifty years, although it was only about level with the Byerley Turk line at the end of the eighteenth century. Of the potent male lines of the present day, the Blandford, the Phalaris, the Teddy, The Boss, the Son-in-Law, the Gainsborough and the St. Simon are all derived from the Darley Arabian. Foaled in 1700, the Darley Arabian was sent home as a four-year-old by Thomas Darley, described as an agent in merchandise abroad and also the British Consul in Aleppo, to his brother Richard in Yorkshire. An elegant bay with a white blaze and three white feet, the horse was stated to belong to 'the most esteemed race among the Arabs both by sire and dam'. The Darley Arabian lived to the great age of thirty and, although he covered few mares except Darley's, his importance is incalculable.

The origins of the Byerley Turk were more obscure. The first of the three great sires to be imported, he may have been a spoil of war and became Colonel Byerley's charger in Ireland. His turn of speed is reputed to have enabled his owner to evade capture while on a reconnaissance before the Battle of the Boyne in 1690. The General Stud Book stated that he covered few well-bred mares, but he got in Jigg a son good enough to carry on a male line which can be traced down to The Tetrarch and Tourbillon lines of the present century.

The Godolphin Arabian, the last of the three, was a brown horse foaled in 1724, and died at Lord Godolphin's place, Gogmagog near Cambridge, at the age of twenty-nine. Thus he was nearly as long-lived as the Darley Arabian, and stood the same height, 15 hands, though most of his progeny were taller. A mass of legend surrounds his career. None of it has been authenticated, but it forms part of the romantic background to the thoroughbred and as such is worthy of preservation. Legend has it that he was discovered in Paris drawing a water-cart. He is supposed to have been a teaser to Hobgoblin when he first went to stud, and got his first chance to cover a mare when Hobgoblin refused to do so. An even more far-fetched version of the story is that he fought Hobgoblin for the favours of Roxana and, having won the battle, covered the mare. Another legend, which probably had more substance because a cat was depicted in his portraits, was that he had a close friendship with the stable cat, who died of grief after his death.

The most up-to-date version of the origin of the Godolphin Arabian is that he was of the Jilfan blood of the Yemen and was exported via Syria to the Arab stud of the Bey of Tunis. He is believed to have been one of four Arab horses presented by the Bey to the King of France. Three were turned out in the forests of Brittany and the fourth was bought by Mr. Edward Coke, of Longford Hall, Derbyshire. He later became the property of Lord Godolphin and was called the Godolphin Arabian.*

Whatever degree of truth may be behind the legends, there is no doubt of his profound influence on the formation of the thoroughbred. The editors of the General Stud Book were able to state in 1891 that there 'is not a superior horse now on the Turf, without a cross of the Godolphin Arabian, nor has there been for many years past'. He got Regulus, the best horse of his day and the maternal grandsire of Eclipse, and his matings with Roxana also had fateful consequences. The first time he covered Roxana he got Lath, who turned out a first-class racehorse and sire; and the second time Cade, who was nothing like such a good racehorse but helped to perpetuate the dynasty which is represented by the Hurry On – Precipitation line of the present time.

### *The Foundation Mares*

The identity of the mares who were the foundation stock of the thoroughbred has been a controversial subject. The General Stud Book† listed seventy-eight of the earliest-known mares, but the list has since been shown to have contained many duplications.

In the 1890's Bruce Lowe traced all mares in the current volume of the General Stud Book to about 50 original mares, but subsequent research, taking into account erosion of a large number of the Bruce Lowe families, has indicated that many fewer foundation mares are represented among modern thoroughbreds by descendants in the direct female line. Cedric Borgnis has expressed the opinion‡ that modern thoroughbreds may be traced back effectively to no more than a dozen foundation

* Sir Mordaunt Milner: *The British Racehorse*, Volume XIV, No. 2, 1962.

† Vol. 1, 5th Edition

‡ *The British Racehorse*, Volume XXV, No. 4, 1973.

mares, of whom the Layton Barb Mare, the Burton Barb Mare, Tregonwell's Natural Barb Mare, Old Bald Peg and Lord D'Arcy's Black-Legged Royal Mare were among the most important. The last named was incorrectly described, because she was really by the Duke of Rutland's horse Blacklegs out of Old Royal, or a daughter of that mare.

The problem of identifying the mares is complicated by the seventeenth and early eighteenth century practice of calling mares after their sires. Thus the Layton and Burton Barb Mares were probably not imported mares at all, but were so-called merely because they were by the Layton and Burton Barbs respectively. After much deliberation Borgnis concluded that the Layton Barb Mare family incorporated ten and the Burton Barb Mare family incorporated four of the Bruce Lowe families.

Some of the original mares were described as 'Royal Mares', and were stated by the authors of the General Stud Book to have been foreign mares imported by Charles II for breeding. This account of them has never been properly authenticated. An alternative and more plausible explanation of the term is that they were mares used by James D'Arcy, who was appointed Master of the Royal Stud at the Restoration, to fulfil his contract to supply the king with twelve 'extraordinary good colts' annually from his own stock. Charles II bred few racehorses himself.

The origins of the thoroughbred foundation mares are wrapped in mystery, and most of them seem to have been of mixed pedigree.

# 2

# *The First Great Racehorse—Flying Childers*

FLYING CHILDERS, FOALED IN 1715, was the first notable landmark in the progress of the thoroughbred, demonstrating that the efforts of breeders to develop a superior breed of racehorses were bringing results. Sportsmen of his day considered him to be 'the fleetest horse that ever ran at Newmarket, or, as generally believed, was ever bred in the world'. Few of his racing performances were recorded, though it is known that he started several times at Newmarket against the best horses of his time, and was never beaten. He beat the Duke of Bolton's Speedwell in a match over four miles in April 1721, and the Earl of Drogheda's Chaunter in a match over six miles in October of the following year. By then he had achieved such a reputation for invincibility that he received forfeit in most of his later matches.

Highly coloured accounts of his prowess passed into the legends of the Turf. He was said to have done a long jump of ten yards with a man in the saddle. When he won a Trial Match against Almanzor and Brown Betty, to each of whom he was conceding a stone, over the Round Course at Newmarket he was timed by the Dukes of Rutland and Devonshire to have covered the 3 miles, 6 furlongs and 93 yards in 6 minutes and 40 seconds, though this was not imagined to be the shortest time in which he was capable of covering the distance. On another occasion he was timed to cover the Beacon Course of 4 miles, 1 furlong and 138 yards in 7 minutes and 30 seconds. These times represent average speeds of 33 and 34 miles an hour, and their credibility must be judged in the light of the fact that the average speed of Birthday Present when he set up the record for 2¾ miles in the Brown Jack Stakes in July 1958 was practically the same.

It was thought that Flying Childers could cover 82½ feet in a second, which represents a speed of 56 miles an hour; yet Indigenous averaged only 42 miles an hour when he set up the world record for 5 furlongs in the Tadworth Handicap at Epsom in June 1960. If the times credited to Flying Childers were correct, the thoroughbred has deteriorated sadly during the last two hundred years, but the lack of accurate timing devices in that era means that they were worthless.

A better idea of the superiority of Flying Childers may be obtained from the result of his trial with Fox in May 1722. Fox was one of the leading racehorses at that time. He won three King's Plates, the Ladies' Plate at York and several matches including one for 2,000 guineas against Lord Hilsborough's Witty gelding. Yet Flying Childers gave him a stone and beat him by nearly two furlongs over an unspecified distance. Flying Childers was indeed a phenomenon.

If Wootton portrayed him accurately, Flying Childers was very much on the leg and appallingly straight in front, with wretched bone and long, weak forearms and second thighs. Probably no more credence should be given to Wootton's portrayal than to the Duke of Devonshire's watch-keeping. He was a bay horse, with a white blaze and four white feet.

Although the validity of contemporary accounts of the speed and conformation of Flying Childers has been impugned, there is no doubt that he marked a significant step forward on the road of thoroughbred progress. For this reason his pedigree should yield vital evidence about the methods of the early breeders. Bred by Mr. Leonard Childers of Carr House, near Doncaster, and sold to the Duke of Devonshire as a young horse, he was described as by the Darley Arabian out of Betty Leedes by Old Careless (by Spanker out of a Barb mare), out of Cream Cheeks by the Leedes Arabian, out of a mare by Spanker, out of the Old Morocco mare, who was Spanker's own dam. The Old Morocco mare was also known as Old Peg, and was by Lord Fairfax's Morocco Barb out of Old Bald Peg, by an Arabian out of a Barb mare. Spanker himself was by the Darcy Yellow Turk.*

It is right to add that the validity of this General Stud Book version of the pedigree of Flying Childers was doubted by C. M. Prior, who discovered another version in the private stud book of the contemporary breeder Cuthbert Routh:

'Childers was got by Darle's Arabian, his dam by Careless,

* See pedigree, page 162.

his granddam by Leedes Arabian, and out of a famous roan mare of Sir Marmaduke Wyvill's. Careless was got by Spanker.'*

Prior was prejudiced in favour of the Routh version by his conviction that it was more than unlikely that 'the ordinary canons of affinity had been disregarded by so eminent a breeder as Mr. Leedes'. This argument cannot be accepted, because in-breeding as close as that found in the official version of the pedigree of Flying Childers was common practice among breeders seeking to improve farm stock during the 18th century. There is no reason why racehorse breeders should not have anticipated their methods.

There is more weight in Prior's argument that Routh's version was contemporary, whereas the official version was taken from Cheny's Racing Calendar for 1743, two years after the death of Flying Childers. On the other hand Robertson, who also made a close study of the early thoroughbred pedigrees, stated that he had 'not the slightest hesitation in accepting the Stud Book and many old pedigree versions of the alliance as correct'.†

If the Stud Book version is accepted, two aspects of the pedigree of Flying Childers are enlightening. The first is that the early breeders were prepared to go to great lengths to perpetuate characteristics they considered desirable; and as Spanker was the best horse to run at Newmarket during the reign of Charles II, clearly they made racing ability the principal standard of selection. The second is that the pedigree of Flying Childers may be traced to exclusively Eastern sources—Arabian, Turk and Barb—within six generations.

Breeders might have been expected to fall over each other to send mares to Flying Childers when he went to stud, but this was not the case. Pick's statement that 'he covered only a few mares except the Duke of Devonshire's' is surprising, since Pick adds the qualification that he was 'allowed by breeders to be a very valuable stallion'. The remoteness of Chatsworth from the nerve centres of racehorse-breeding in Yorkshire may have had a good deal to do with this otherwise culpable neglect.

Pick's opinion of Flying Childers as a stallion was not shared by Colonel Gilbert Ironside, who emerged briefly from the obscurity that shrouded his military career to contribute 'A Dissertation on Horses' to the supplement to the General Stud Book published in 1800. Ironside wrote: 'It is remarkable that

* The Royal Studs of the Sixteenth and Seventeenth Centuries.

† The Lonsdale Library Flat Racing.

scarcely any fleet coursers sprang from the celebrated Childers, though himself the swiftest horse ever known; on the contrary his race have proved eminently defective; declining, like our modern nobility, with an uncommon precipitation of degeneracy.' It is true that Flying Childers did not establish an enduring male line. This achievement was reserved for his, from the racing point of view, worthless full-brother Bartlett's Childers. This horse was never trained because he was a blood-vessel breaker. At first he was known by the descriptive name of Bleeding Childers, but this must have been regarded as bad publicity when he went to stud and was changed to Bartlett's Childers, after his owner, Mr. Bartlett, of Masham in Yorkshire. In spite of his disability, he sired so many good horses that he became one of the leading stallions of his day. He was the great-grandsire of Eclipse.

On the other hand, the failure to found a male line does not make Flying Childers a failure altogether. Indeed Colonel Ironside's condemnation was completely unjust. One of Flying Childers's sons was Snip, an indifferent racehorse but a successful sire who got Snap, a horse allowed to be equally as good a runner, if not superior, to any of his time. Snap had a tremendous influence as a sire of brood mares, although his daughters were inclined to be bad-tempered. He sired the dams of the Derby winners Saltram and Sir Peter Teazle, besides appearing three times close up in the pedigrees of the Derby winners Whalebone, Whisker and Phantom and twice close up in the pedigree of the Derby winner Pope. Breeders had a saying: 'Snap for speed and Matchem for truth and daylight'. On the final count it was Colonel Ironside, not Flying Childers, that was discredited.

### *The Second Great Racehorse—Eclipse*

'Eclipse was never beat, and was allowed by all ranks of sportsmen to be the fleetest horse that ever ran in England, since the time of Childers', wrote Whyte in his *History of the British Turf.* The reference was to Flying Childers, and the link between the first two great English racehorses was Bartlett's Childers, the full-brother of Flying Childers and the great-grandsire of Eclipse.

The career of Eclipse summed up all the romantic elements that have been the making of Turf history. Unlike Flying Childers, he did not depend for his reputation mainly on form shown in trials. He was fully tested on the racecourse and did not merely win all his races—defeat for him seemed inconceivable. As to his breeding, he had a pedigree in which luck had played its part at every possible turn. Alone among the thoroughbreds of all time, he has added a phrase to the English language: 'Eclipse first, and the rest nowhere.'

Eclipse was bred by William, Duke of Cumberland, who had both his sire Marske and his dam Spiletta in his stud in Windsor Forest. Cumberland also bred Herod, known alternatively as King Herod, and the Eclipse – Herod cross was to prove one of the most successful in the whole development of the thoroughbred. Thus the man who conducted with such ruthless efficiency the campaign that led to the annihilation of the clansmen at Culloden made another, more peaceful contribution to his own and later times.

Eclipse was foaled in 1764, the year of the Great Eclipse, after which he was named. Cumberland did not live to see the racing career of this phenomenon that he had bred, for he died in October of the following year. Eclipse was then put up for sale and bought by William Wildman, a Smithfield meat salesman, for 75 guineas. The colt had been knocked down for 70 guineas a few minutes before Wildman's arrival, but, on consulting his watch, Wildman discovered that the sale had taken place before the advertised time, and successfully demanded a resale. This was the first remarkable twist to the career of Eclipse.

Before long Eclipse began to show the bad temper, or what is sometimes euphemistically called the high courage, which has characterised a good many outstanding racehorses. No doubt the heightened vitality that helps to provide their speed and sustains their spirit makes them impatient of any kind of restraint. Eclipse was so difficult to manage that Wildman was in two minds whether to have him gelded. Fateful decisions often hang by a slender threat. Instead he was sent to a rough-rider near Epsom, who combined poaching with his legitimate business. Besides his normal exercise by day, Eclipse frequently carried the rough-rider on poaching expeditions by night, and this strenuous way of life not only kept his spirits in check but helped to develop muscle and sinew.

When Eclipse went into serious training in preparation for his first race in May as a five-year-old his extraordinary quality soon became apparent. The word went round, and some of the touts travelled post-haste to Epsom Downs in an attempt to watch one of his trials. They arrived too late, but they found an old woman who told them that she had just seen 'a horse with white legs running away at a monstrous rate, and another horse a great way behind, trying to run after him; but she was sure he would never catch the white-legged horse if he ran to the world's end'.* Eclipse really had white on only one of his legs, and this consisted of a stocking on his off hind. He also had a white blaze. His first race was at Epsom and was a £50 Plate for horses that had never won £30, matches excepted, run in four-mile heats. Dennis O'Kelly, an Irish adventurer who rose to the rank of lieutenant-colonel in the Westminster Regiment of the Middlesex Militia, betted evens and 6 to 4 he could place all the four runners and, when called upon to declare the order of finishing, made the statement that has become proverbial. His forecast was proved correct when Eclipse, galloping with his head held so low that it seemed to be practically touching the ground, and with his jockey Oakley powerless to hold him, went right away from his three rivals in the last mile of the first heat and distanced them all. A month later O'Kelly bought a half-share in Eclipse for 650 guineas, and in April 1770 bought the remaining half for a further 1,100 guineas. It was a magnificent investment, because O'Kelly reckoned that he netted more than £25,000 from the horse during his racing and stud careers. Altogether Eclipse won eighteen races, including eleven King's Plates, for seven of which he walked over. Oakley and the other two jockeys who rode him in some of his races, Merriott and Fitzpatrick, found that it was useless to attempt to hold him, and simply let him stride along as he pleased. As a result his overwhelming superiority over his contemporaries was never concealed; for instance, when he won the subscription purse at Guildford in August 1770 he was more than a distance ahead of his rivals by the time that he had covered half of the four-mile course.

No racehorse has had his conformation, character and idiosyncrasies as minutely analysed as Eclipse. His fiery temper, his impetuosity and his thick-windedness, which caused him to puff and roar so that he could be heard at a considerable distance, made him as alarming to meet in a confined space as a runaway engine in a tunnel. He was a big horse in every way; he was tall as horses went in the last quarter of the eighteenth century, and was deep-girthed and had a grand middle-piece. For a long time his height at the withers was accepted as 16 hands 2 inches. The

* Youatt: *The Horse.*

latest estimate is that he stood no more than 15 hands 3 inches, which still made him a big horse by the standards of his time.* Some leading horses of the period were much smaller. For instance, Gimcrack, whose excellence is commemorated in the Gimcrack Stakes, stood only 14 hands and $\frac{1}{4}$ inch. One point about him which struck observers particularly was that the highest point of his quarters was about an inch above his withers. He had great hip to hock length and a short, powerful forearm, and this combination gave tremendous elasticity to his stride. He had perfectly placed, sloping shoulders. These qualities made him the most efficient galloping machine yet produced.

Granted that Eclipse was a superb racehorse, his pedigree makes a most unsatisfactory field for study. In the male line he traced back through Marske and Squirt to Bartlett's Childers, and not one of these measured up to the standard expected of the ancestors of a horse who helped to raise the whole quality of performance of the thoroughbred to a new high level. Bartlett's Childers, as we have seen, was a blood-vessel breaker who could not be trained. Squirt won six races without gaining the reputation of being anything out of the ordinary. Probably he was troubled by laminitis when he was in training; while he was at stud the condition became so bad that he was ordered to be shot, but was reprieved on the intervention of his groom. No doubt on account of his unsoundness, he covered few well-bred mares. Marske was about the best of his progeny and won a Jockey Club Plate of 100 guineas at Newmarket, but had his limitations and was beaten in the last two races in which he took part.

Marske was a private stallion in the Duke of Cumberland's stud until the death of the Duke, when he was sent up to the dispersal sale of the Duke's bloodstock at Tattersall's and sold to a farmer for a trifling sum. At that time he was fifteen years old and considered a very indifferent stallion. The next season he covered country mares in Dorset at half a guinea. Afterwards he was bought by William Wildman for 20 guineas. The seller was glad to be rid of the horse at any price, but Wildman must by then have been aware of the exceptional promise shown by Eclipse, and proved for a second time that he knew a good bargain when he saw one. With the racecourse triumphs of Eclipse the reputation of his sire soared, and Wildman passed on Marske to Lord Abingdon for 1,000 guineas. In later years Marske, who lived to the great age of twenty-nine, stood at 100

* G. Rathbone: *The British Racehorse*, Vol. XV, No. 5.

guineas, a fee which was not charged again for the services of any sire for a very long time. He never got another horse of Eclipse's class. To complete the picture, Eclipse's dam Spiletta was beaten the only time she ran, and Snake, to whom he was inbred, never raced.

Perhaps the real heroes among the ancestors of Eclipse lie, hidden and unsung, elsewhere in his pedigree. Hautboy may have been one of them. The Darley and Godolphin Arabians appear only once in the pedigree; Hautboy, by the Darcy White Turk, appears no less than nine times in the first seven generations. This concentration provides further evidence of the degree of inbreeding to which the early breeders were prepared to resort, but as Hautboy himself was never trained and got no horses known to have possessed exceptional racing ability, it is impossible to assess the value of the inbreeding to him. Regulus, the maternal grandsire of Eclipse, may have been another key influence. Foaled in 1739, Regulus won eight King's Plates and a £50 Plate, was never beaten, and was vastly superior to any other horse of his time. He went to stud near Catterick, where he was an outstanding success. He was leading sire of winners eight times and was also a first-rate sire of broodmares. The granddam of Regulus, one may add, was closely inbred to Hautboy.

### *The Stud Career and Influence of Eclipse*

The lifetime of Eclipse covered a transitional period of Turf history; taking his racing career as one leg and his stud career as the other, he may be said to have straddled two different eras. He himself raced in conditions of heavy weights and long distances that were little changed from the days of Flying Childers; eleven of his victories were gained in King's Plates, races founded by Charles II for the purpose of promoting strength and endurance in the thoroughbred and in which the runners had to carry 12 stone over four miles. He did not run until he was five. His progeny raced mostly in radically changed conditions that resembled those of the twentieth century more than those of the times of Flying Childers. Strength and endurance were no longer the ideals for breeders to aim at; speed and early maturity had taken their places, and the Derby, Oaks and St. Leger, confined to three-year-olds and run over distances of less than two miles, were on their way to acceptance as the supreme tests of merit.

In spite of the revolutionary contrast between the conditions in which he raced and the conditions in which his progeny raced, Eclipse achieved enormous success as a sire. He got three winners of the Derby (Young Eclipse, Saltram and Sergeant), and one of the Oaks (Annette). Altogether he got 344 winners of £158,047, in days when race values were infinitesimal compared with prize-money in the second half of the twentieth century. The large majority of the leading stallions of modern times can be traced back to him in the direct male line. His outstanding merit, as sire besides racehorse, is beyond dispute. All the same, it may not be out of place to enquire exactly how great a sire he really was.

The answer to that question will depend on the importance that is attached to 'lines', whether male or female, in the pedigrees of thoroughbreds. This aspect of breeding will be considered in a later section of the book, but it stands to reason that if a certain pattern of mating is found at the roots of a high proportion of the various branches of the Eclipse male line, then all of the credit cannot be claimed for Eclipse himself.

Three other horses whose careers overlapped that of Eclipse may be placed in the category of great sires. They are Matchem (1748–81), Herod (1758–80) and Highflyer (1774–93). Matchem lived too soon to get many Classic winners. The St. Leger was first run in 1776, the Oaks in 1779 and the Derby in 1780, but he sired one winner of the Oaks (Teetotum) and one of the St. Leger (Hollandaise), and all told 354 winners of £151,097. Herod got three winners of the Oaks (Bridget, Faith and Maid of the Oaks) and one of the St. Leger (Phenomenon), and 497 winners of £201,505 altogether; and Herod's son Highflyer got three winners of the Derby (Noble, Sir Peter Teazle and Skyscraper), one of the Oaks (Volante), four of the St. Leger (Omphale, Cowslip, Spadille and Young Flora), and 470 winners of £170,407 altogether.

Matchem was leading sire of winners three, Herod eight and Highflyer twelve times. Eclipse was never leading sire, though he was often tantalisingly close, and was second no fewer than eleven times—seven times to Herod and four to Highflyer. It is clear that as a stallion Eclipse could claim nothing more than parity with the best of his contemporaries.

A clue to the status of Eclipse may be found in the conditions for the July Stakes at Newmarket. This, the first of the important two-year-old races, was inaugurated in 1786 with the stipulation that the progeny of Eclipse and Highflyer should carry three pounds extra. Apparently he was regarded as a great sire, but no greater than Highflyer. Covering fees may also be a reliable guide to the standing of a sire with breeders. A fee of 50 guineas was charged for the services of Eclipse when he first went to stud, but this was reduced to 25 guineas a year later and never rose above 30 guineas again. By contrast a fee of 15 guineas was charged for Highflyer at first, but this was raised by stages to 50 guineas and reduced to 30 guineas only for the last season before he died.

The main characteristics of the progeny of Eclipse were that they were light-fleshed, excitable and active, and came to hand early. The breeders of the time used to say that 'the Eclipses were speedy and jady, and the Herods hard and stout'.* Stout meant stout-hearted. If this were so, the qualities of the two sires should have been complementary, and the best results obtained from matings which combined them. This indeed was the case, and the pre-eminence of the 3rd and 4th Dukes of Grafton as breeders about the turn of the century was based on constant but judicious repetition of the Eclipse – Herod cross. Waxy, who won the Derby for Sir Ferdinando Poole in 1793, was by Pot-8-Os (by Eclipse) out of Maria by Herod. Afterwards the Duke of Grafton bought Waxy, mated him with the Highflyer mare Prunella, and bred the 1809 Derby winner Pope. By a continuation of the same breeding pattern, the Graftons mated Prunella's daughter Penelope with Waxy again and again and so bred the Derby winners Whalebone and Whisker besides Web, who was sold to Lord Jersey and became the foundation of that breeder's success, producing the Derby winner Middleton.

Dennis O'Kelly was an equally staunch believer in the Eclipse – Herod cross. He bred the 1784 Derby winner Sergeant by sending the Herod mare Aspasia to Eclipse. The value of the Eclipse – Herod cross was demonstrated time and again in the breeding of Derby winners round about the turn of the century. Skyscraper, the winner in 1789, was by Highflyer out of Everlasting by Eclipse, and John Bull, the winner three years later, was by another of Herod's sons, Fortitude, out of Xantippe by Eclipse. The 1795 winner Spread Eagle was by Volunteer (by Eclipse) out of a Highflyer mare, the 1799 winner Archduke was by Sir Peter (by Highflyer) out of Horatio by Eclipse, and in the following year the winner Champion was by Pot-8-Os out of Huncamunca by Highflyer.

* 'The Druid': *Silk and Scarlet.*

Waxy represented the Eclipse – Herod cross in its highest form. Although he was a bay whereas his grandsire was a chestnut, he had Eclipse's white stocking on his off hind. He was a horse of superb quality, with a beautifully chiselled Arab-like profile, and his supreme asset as a sire was that he passed on his quality to most of his progeny. 'The Druid' described him as 'the modern ace of trumps in the stud book'. Waxy was the principal agent in perpetuating the Eclipse male line. Subsidiary branches of the line were founded by Eclipse's sons Joe Andrews and King Fergus, and each of these branches has extended generation by generation to reach well into the twentieth century. In the case of Joe Andrews the next link in the chain of heredity was Dick Andrews, and in the case of King Fergus the next link was Hambletonian. Dick Andrews and Hambletonian were both out of Highflyer mares.

These facts suggest that the reputation of Eclipse is due for reappraisal. As a sire of winners he was no better than three other horses who were at stud at the same time, and the earnings of his progeny were less than those of the progeny of Herod and his son Highflyer. In the wider context of thoroughbred development, the prestige of Eclipse has been artificially inflated by too much emphasis on 'lines'. What Eclipse accomplished in this respect was the result of teamwork, or what is usually called a 'nick' with Herod, whose 'line' has not ramified to the same extent or achieved anything like the same numerical strength. Eclipse and Herod together, not Eclipse alone, were responsible for a decisive advance in the development of the thoroughbred. Eclipse ought not to be pushed off the pedestal that he has occupied for so long, but room must be made for Herod to stand beside him.

### *Was there an 'X' Factor?*

The period of predominance of the Eclipse – Herod cross in the first quarter of the nineteenth century marked the end of the formative stage of the thoroughbred. After that period there was to be no further spectacular advance, whereby the whole standard of performance was raised to a higher plane within a single generation, at least until the St. Simon era near the end of the century. The thoroughbred had emerged as a distinct breed incapable of further improvement by fresh crosses with the Arab and other foreign stock in which it had originated.

In 1850, when the development of the breed had been in

progress for roughly two hundred years, Captain (afterwards Admiral) Rous gave a summing-up of the situation which may be conveyed most easily by a series of brief quotations from his work, *On the Laws and Practice of Horse Racing*:

'It is generally believed by the most learned men of the turf that a first-class English racehorse would give 6 st. to the best Arabian that can be found, for any distance under ten miles.

'The clearest proof of the improvement which has taken place in the English racehorse is the fact that no first or second cross from the imported Arab, with the exception of the produce of one mare by the Wellesley Arabian (Fair Ellen), is good enough to win a £50 plate in the present day; whereas in 1740 our best horses were the second and third crosses from the original stock; and we have no reason to assume that the Arabian horse of 1850 has degenerated from his ancestory of 1730.

'I . . . suspect that the form of the best racehorse of 1750 is inferior to that of a common plater of the present day.

'The English racehorse boasts of a pure descent from the Arabian, and under whatever denomination the original stock of our thoroughbred horses have been imported . . . they were considered as the true sons and daughters of the desert.'

Statements by Admiral Rous have usually been allowed the weight of *ex cathedra* utterances, but before they can be accepted now must be submitted to the test of any additional evidence that has come to light since his day. This evidence, such as the form of St. Simon compared with that of Arabs, has served to substantiate the Admiral's case in respect of improvement from the original stock, as will be shown at a later stage of the book. On the other hand, the composition of the original stock has remained a controversial question.

Reference has already been made to the fact that the appearance of certain modern thoroughbreds contains an element of coarseness that suggests a non-Arab source. A further doubt is sown by the discovery that few of the Arab and other Eastern horses imported during the hundred years following the middle of the seventeenth century were used for racing, presumably because they were not good enough to stand a chance of winning. Even in the early days the imported horses were seldom recorded as winners. William III's Turk beat Lord Carlisle's Spot in a match for £500 over four miles at Newmarket in April 1698, when it was said that 'the swells lost a world of money on Spot', and the Prince Consort's grey Barb beat Frampton's

Thiller in another match at Newmarket seven years later. Otherwise the imported horses seem to have made a poor showing. Of the famous members of the foundation stock, only the Byerley Turk is reported to have given positive evidence of speed, and that was in the unauthenticated incident of the reconnaissance before the Battle of the Boyne. Is it therefore right to deduce that the ancestry of the thoroughbred contained an 'X', or untraced speed factor, independent of the importations of the seventeenth and eighteenth centuries, and with which the qualities of those imported stallions and mares had to be blended before the thoroughbred could evolve? Was there, in short, a native English element in the making of the thoroughbred?

Modern research indicates an affirmative answer to these questions, provided that the adjective 'native' is given broad interpretation. Racing had been an established sport in the British Isles for centuries before the Restoration, and the stock from which racehorses were bred had been refreshed continuously by successive importations from many different sources. As a result the stock had become so heterogeneous by the first half of the sixteenth century that breeding to race was virtually impossible as a planned operation.

The prime need was the introduction of stock of greater genetic purity which would provide the basis for rational breeding policies. And breeders turned naturally to Arabian and related Eastern breeds, which were the most genetically pure available, for the purpose. The Eastern horses possessed exquisite quality and the ability to breed true to type; grafted onto the stock already existing in the British Isles, they provided the conditions essential for a determined and constantly maintained attempt to improve racing performance over many equine generations.

The General Stud Book version of the pedigree of Flying Childers may have been authentic, but Rous was wrong in thinking that all thoroughbreds were true sons and daughters of the desert. Nevertheless he was certainly correct in pointing out the great scope for improvement in size and performance inherent in the original stock. As he observed justly: 'The change of climate, the pasture, and extreme care and attention in breeding by the best stallions have wonderfully increased their size, their strength and their endurance.'

# 3

# *Heredity and Environment*

THE YEAR 1850 WAS A CONVENIENT milestone in thoroughbred history. It saw the publication of Rous's trenchant remarks; it marked the end of two centuries of evolution and a time when it had become absolutely certain that the thoroughbred was a distinct breed of racehorse incapable of improvement by further admixtures of the original stock. Nevertheless progress did not come to a dead stop. The dramatic improvement of the early days was not to be repeated, but there is little doubt that the thoroughbred of 1950 was superior, at least as far as speed is concerned, to the thoroughbred of 1850.

Whether the breed was advancing on a broad front or not, competition between breeders to make the best of their resources and produce the outstanding members of each generation was intense. This competition has generated scores of breeding theories and mating systems with the dual purpose of producing better horses and reducing the area within which chance is able to operate. Many of these theories have been based on painstaking research, but surprisingly few efforts have been made to relate them to the progress of scientific knowledge.

It should be obvious that any attempt to evaluate the various breeding theories and mating systems must have as its starting point the best available information about the manner in which qualities are transmitted from one generation to another. It is in order to provide this essential criterion that an account of the mechanics of inheritance is now given.

### *Heredity and Environment*

Two factors determine the performance of the thoroughbred on the racecourse and at stud—heredity and environment. The word heredity describes that part of its make-up that it has received from its ancestors, and may be compared to the foundations of a building. The foundations determine the general shape, appearance and proportions of the building, and even its height and weight, for if these exceeded certain limits the foundations would not be able to uphold them. Within the limits imposed by the layout and strength of the foundations there is tremendous scope for the architect and the decorator, and these aspects of the building correspond to the influence of environment.

It is tempting to try to assign relative degrees of importance to heredity and environment, but the attempt is foredoomed to be fruitless. Both are essential to the well-being, the development, the very existence of the animal. Environmental influences affect development from the moment that the female egg is fertilised by the male sperm, and the fertilised egg, the first cell of the new organism, may not survive the first minutes if harmful micro-organisms are present in the womb. The feeding and health of the mare during pregnancy, the conditions of foaling, the milk supply of the mare and food intake after weaning, climatic conditions throughout life, the care with which it is handled as foal and yearling at stud, the skill of trainer and lad in racing stables, the incidence of disease and injury, the severity or otherwise of its races, and the quality of handling and feeding when it is back at stud again—these are some of the environmental factors that may influence its career, yet without being able to alter its identity as a horse. Peas grown in poor soil and a period of drought may be stunted and tasteless, but they will still be peas, not beans or lentils. Similarly a thoroughbred reared in the least favourable environment will still be a thoroughbred, if a poor specimen of the breed, not a hackney or Shetland pony.

### *Hereditary Factors*

The hereditary factors that determine physical, mental and nervous constitution are transmitted by means of microscopic chemical substances called genes, which are the units of inheritance. Each gene is responsible for a separate characteristic, or 'character', as it will be called hereafter. The science of the inheritance of characters is called genetics, and the whole inherited make-up, or genetic constitution, of an individual is known as its 'genotype'. A genotype consists of thousands of genes.

Genes are strung together like beads on thread-like bodies

called chromosomes, and chromosomes are paired off and contained in cells, which are the smallest vital elements of an organism. The first cell of a new organism that is to become a thoroughbred is formed when the female egg is fertilised by the male sperm, and contains the whole complement of chromosomes and their attendant genes which compose its inheritance.

The new organism is built up from this tiny beginning. The first cell divides to form two cells, those two cells divide again, and so the number of cells continues to grow by geometrical progression for a long time, though eventually some of the cells cease to divide, temporarily or permanently. In this way the body of the animal is developed by a system of dividing to multiply.

This system of division does not resemble the division of a sum of money into two parts, each part containing half the pounds and pence of the original sum. Cell division results in the formation of two identical cells containing the same team of chromosomes and genes as the parent cell. This is made possible by the duplication of the chromosome population of the parent cell before division, and one of the parts so formed goes to each of the new cells. In this way every unit of the animal's body contains the whole complex of factors which determine that it shall be a thoroughbred and no other animal.

When the animal reaches sexual maturity it begins to form sex cells as well as body cells. These differ from the body cells because they are the result of two cell divisions in which the chromosomes split only once. The effect is to reduce the number of chromosomes so that the sex cells contain only half as many chromosomes as the ordinary body cells, but this reduced number comprises one chromosome of each of the original chromosome pairs. The identical process occurs in the formation of male and female cells. Thus when at fertilisation the sperm and the egg fuse and the original number of chromosomes is restored, the first cell of the foal-to-be contains two sets of chromosomes, one of paternal, the other of maternal origin. In this cell the chromosomes of the same type pair off and lie alongside each other.

### *Inheritance of Characters*

In a breed like the thoroughbred all the specimens bear sufficient resemblance to each other to convey the fact that many genes must be common to all. On the other hand, there is also too much variety in looks and performance within the breed to be accounted for by differences of environment, so the genotypes must differ in many respects. Pinza and Hyperion were roughly equal in racing ability and both won the Derby, and an observer would not hesitate to identify them as members of the same breed; yet Hyperion was chestnut, 15 hands 1½ inches tall and of the most exquisite quality, whereas Pinza was bay, 16 hands 1 inch tall and of coarser fibre. It is also possible for a Derby winner and a selling plater, representing the two extremes of racing class, to look alike.

Thus there is great variety within the breed as a whole. It is also clear that an individual thoroughbred is liable to receive a variety of characters from its parents. For example, of a pair of colts by Big Game out of Cap D'Or, one, Makarpura, was chestnut and stayed only seven furlongs, and the other, Michelino, was bay and stayed two miles.

These two-fold variations lead to the heart of the whole problem of the inheritance of characters, and the solution lies in the laws which govern the behaviour of genes. In the thoroughbred each cell contains thirty pairs of chromosomes, and each chromosome has many genes strung upon it in single file. These genes are arranged on the chromosomes in fixed order so that the genes for any particular character are opposite each other in the chromosome pairs. The two genes of a pair, however, may not be identical. Taking colour as an example, one gene may be the factor for chestnut where as its opposite number on the companion chromosome may be the factor for bay, bay and brown being genetically similar. There are likely to be other gene differences in the same chromsome pair and also in the other twenty-nine pairs that make up the genotype.

At the formation of the sex cells pure chance decides which member of a chromosome pair shall be taken for each cell. Thus, assuming at least one gene difference in each chromosome pair, a single horse may produce a total of 2 × 30, or 60 different kinds of sex cell. The identical process takes place in the formation of male and female sex cells and it is again chance that decides which male and female cells unite at coition, so the potential variability, even between brothers and sisters, is very great indeed.

This explains how own brothers and sisters may differ as much as Makarpura and Michelino. Even when own brothers resemble each other as closely in aptitude as Dante and Saya-

*(1)* Roxana (from a mezzotint by Seymour). Foaled in 1718, Roxana was one of the most famous of eighteenth-century brood mares

*(2)* Eclipse. Foaled in 1764, the year of the great eclipse, Eclipse was the second great racehorse and one of the most influential sires in thoroughbred history

*(3)* Flying Childers and Dimple. Flying Childers, foaled in 1715, was the first great racehorse. Dimple, foaled about 1708, was closely related to the dam of Flying Childers and sired a number of winners

*(4)* The Godolphin Arabian. He was imported into England in 1730 and became one of the most famous progenitors of the thoroughbred. He died in 1753, when he was supposed to be in his twenty-ninth year

*(5)* The Flying Dutchman (from a portrait by Maggs). The winner of the Derby and St Leger in 1849, The Flying Dutchman was the maternal grand-sire of Galopin, the sire of St Simon.

*(6)* Gincrack (from a mezzotint by Bradford) Although he stood only 14 hands $\frac{1}{4}$ inch, Gimcrack was one of the best racehorses in the eighteenth century

*(7)* Stockwell. Foaled in 1849, and winner of the 2,000 Guineas and the St Leger, he was afterwards known as 'The Emperor of Stallions'

*(8)* St. Simon. Foaled in 1881, he was probably the greatest racehorse and sire the world has ever seen

jirao, who won the Derby and the St. Leger respectively, further repetitions of the same mating are not guaranteed to continue to produce high class offspring. Indeed subsequent matings of Nearco and Rosy Legend, the parents of Dante and Sayajirao, produced three more live foals, but only one of these, Dark Rose, was a winner, and her solitary success was gained in a mile race worth £276.

A corollary is that a breeder, having decided that there are cogent reasons for sending a mare to a particular stallion, ought to persevere with that mating in the face of early disappointment because the potential of it may not be realised at the first, or even at the second attempt. No breeder has reaped fairer rewards for remaining true to his convictions that Captain John Gubbins, who mated his mare Morganette with Kendal five times before he obtained Galtee More, the Triple Crown winner of 1897; and then switched her to St. Florian, and repeated that mating four times before he obtained Ard Patrick, who won the Derby in 1902 and beat Sceptre in an epic struggle for the Eclipse Stakes the following year.

This separation into different sex cells, and thence into different offspring, of the two members of any pair of contrasting genes possessed by an individual, is called segregation. Obviously segregation is of the greatest importance in the processes of heredity.

### *Sex Determination*

One of the chromosome pairs determines sex and the characters directly connected with sex. The so-called X chromosome is the factor for female sex, and the so-called Y chromosome is the factor for male sex. Y dominates X. Every egg is equipped with an X chromosome, but a sperm may have an X or a Y chromosome, with equal chances as to which it may be. If a Y chromosome penetrates the egg, a colt will result, and if an X chromosome penetrates the egg, a filly will result.

The number of colts and the number of fillies produced by this process should be roughly equal, and of the live births recorded in Volume XXXVI of the General Stud Book 12,497 were colts and 12,411 were fillies.

In some creatures, notably poultry, genes controlling other characters not directly connected with sex are carried by the sex chromosomes. No important sex-linked genes have yet been clearly identified in the thoroughbred, though there is some evidence that the inheritance of haemophilia, which is extremely rare, and certain colour patterns may be sex-linked.

*Additional Factors Affecting Variability*

Three additional factors have an impact on variability. The first is the phenomenon known as 'crossing over', which occurs during the double cell division at the formation of the sex cells. In the turmoil that attends this process the two strands of the chromosomes sometimes become entwined and break as the strings do in the game of conkers; but whereas in conkers a deliberate re-tying of the strings is necessary before the game can be resumed, the broken chromosomes rejoin automatically, though often in the reverse order. In other words, the pieces 'cross-over', and a piece of the maternal chromosome becomes attached to the paternal chromosome, and vice versa.

This 'crossing-over' transforms the normal processes of heredity, as it leads to a mingling of the paternal and maternal genes, and the chromosomes which are passed on to the offspring are different from those inherited from the parents.

The second factor is mutation, which denotes a spontaneous change which may affect a gene and the character controlled by it. Important mutations may alter the entire course of evolution of a species from time to time, but they occur infrequently and are unlikely to upset the plans of any individual breeder of thoroughbreds.

Crossing-over and mutation tend to increase variability. On the other hand the third factor, which is the linking together of genes in groups, tends to limit variability. This linkage results from the fact that genes determining certain characters are always attached to the same chromosomes and therefore are passed from one generation to another as a unit. These groups can be broken up only by crossing-over. Genes situated close together on a chromosome are obviously less likely to be split up by this process than those that are far apart on the string.

*Breeding Pure and Dominance*

It is evident from the foregoing that a thoroughbred may carry in his body cells any particular gene in double strength, meaning that the gene concerned appears on both chromosomes of a pair, or in single strength, meaning that the gene concerned appears on only one chromosome of a pair. If he has the gene in double strength it goes into all his sex cells and is transmitted to all his progeny, and he may be said to breed pure for the character controlled by that gene; if he has the gene in single strength it will go into only half his sex cells and be transmitted to only half of his progeny, and he may be said to breed impure for the character controlled by that gene.

When two genes of different kinds are paired in an individual the resulting character may be intermediate and represent a blend of the two. It is more usual for one gene character to dominate the other and make its presence felt exclusively. This gene is known as the 'dominant gene'; the gene which is dominated is known as the 'recessive gene', indicating that it is not destroyed but merely pushed into the background, to reappear perhaps in future generations when circumstances favour it.

*Mendelism*

Some of the most important discoveries about heredity in general and the behaviour of dominant and recessive genes in particular were made by Gregor Mendel, the abbot of an Augustinian monastery in what is now Czechoslovakia. Mendel published his findings, based on experiments with flowers and vegetables, in a paper read to the Brno Natural History Society in 1865, but their significance was not recognised by scientists elsewhere for a long time. It was not until the beginning of this century that Mendel's ideas were rediscovered and became the foundation of the science of genetics. By the time that 'Mendelism' had been thoroughly digested many theories about the breeding of thoroughbreds had been formulated which would have been regarded as scientifically untenable if it had won immediate acceptance.

The character which illustrates the behaviour of dominant and recessive genes most clearly is colour, since it is determined by a single gene, in connection with certain diluting genes, in the thoroughbred. Mendel found that some sweet peas bred pure for red colour and others for white. When he crossed a red with a white he obtained all red offspring, but when he mated these cross-breds together he got offspring in the proportion of three reds to one white. Further experiments revealed that of these cross-breds one was a pure-breeding red, two were impure-breeding reds, and one was a pure-breeding white. Thus the red dominated white, and whenever red and white were found in combination, the red appeared and the white was thrust into the background; and as long as this cross-breeding was persisted in, the ratio of three reds to one white was maintained.

What holds good for plants holds good for thoroughbred horses, in which bay is dominant to chestnut. For simplicity's sake the far less numerous grey horses have been excluded from this survey, though grey colour also is inherited in obedience to Mendelian principles and is the property of a dominant gene. The ratio of three bays or browns to one chestnut is constant and the larger the same taken the more closely will the numbers approach this exact ratio. A random sample of 100 mares taken from Volume XXXIV of the General Stud Book produced a score of 71 bays or browns, 25 chestnuts and 4 greys.

The point is crucial, and to grasp it it is essential to understand how the Mendelian proportions come about. The gene for bay colour may be represented by B, and the gene for chestnut by C. Thus the genes in the chromosome pairs of the body cells of a pure-breeding bay may be expressed as BB, and the genes in the chromosome pairs of the body cells of a chestnut as CC. At the formation of the sex cells, each sex cell of the pure-breeding bay will contain B, and each sex cell of the chestnut will contain C. When the pure-breeding bay is mated with the chestnut the body cells of all the offspring will contain BC, or one gene for the bay and one for chestnut. All the offspring will be bay because bay dominates chestnut, but they will all be impure-breeding bays. When these inpure-breeding bays form their sex cells, half the cells will be B and the other half C, and it follows that the mating together of these impure-breeding bays will produce offspring in this ratio: one BB, two BCs and one CC, which is one pure-breeding bay, two impure-breeding bays and one chestnut.

To sum up, a pure-breeding bay will have none but bay offspring unless mated with a grey in which case the offspring may be grey; matings of impure-breeding bays with other impure-breeding bays will result in bay and chestnut offspring in Mendelian proportions. A chestnut mated with a chestnut can have only chestnut offspring, chestnut colour being the property of a recessive gene. Every grey foal must have at least one grey parent because the gene controlling this colour dominates all other colour genes and expresses itself wherever it is present. It is of course impossible to tell a pure from an impure-breeding bay by looking at them, a fact which indicates the acuteness of the problem which faces the thoroughbred breeder when he is dealing with characters more complex and less easily discernible than colour.

Other easily observable kinds of dominance occur in ears, in which pricked are dominant to lop ears, and profiles, in which concave, or Arab type, is dominant to convex, or roman-nosed, type.

What applies to a simple character like colour applies equally to other qualities of the thoroughbred which may be determined not by a single gene, but by a combination of genes. These qualities, which include such vital attributes as speed and stamina, involve far more complicated problems for the breeder. Anyone who wished to produce a breed of chestnut racehorses could do so in a single generation by using only chestnut stock; it is much harder to produce a breed of pure-breeding bays, since any bay horse must have a large number of offspring before there is a reasonable assurance that he does breed pure for that character. When several genes affect the quality with which the breeder is concerned, the elimination of an unwanted recessive gene can be at best a long-term proposition.

### *Selection and the Interaction of Heredity and Environment*

Selection occurs on every occasion that one rather than another of two or more possible matings takes place. When a breeding theory is applied, selection represents the point at which theory is translated into practice.

In nature selection is a harsh matter and only the fittest survive. In a herd of wild horses the strongest and most virile of the stallions is king and covers the mares until one younger and more vigorous than he comes along to depose him. No mare lives long enough to produce foals unless she has the constitution to survive in natural conditions. Weakly members of both sexes are eliminated ruthlessly.

Selection among domesticated animals is different because individuals possessing defects which would render survival impossible in natural conditions can be preserved for breeding. The breeder has in his mind's eye an ideal offspring, or has an idea of the direction in which he would like to make progress, and exercises his judgement to choose between the various available matings accordingly. His judgement has to replace the disciplines of natural selection.

Some aspects of selection will be discussed in a later part of the book. At this stage it is necessary only to mention the way in which successful selection may depend upon the interplay of heredity and environment. As environment determines the degree to which a character is developed by the animal, selection for any given character can be made only under conditions favourable for the development of that character. This explains why some stallions may fail in one country but achieve success when they are transferred to another. For example the first Derby winner Diomed was an almost complete failure at stud in England, and was exported to Virginia at the age of 21 for the derisory price of 50 guineas. In Virginia he thrived marvellously, lived for another ten years, sired one high class horse after another and founded a male line of decisive importance for the development of the American thoroughbred. He was the great-great-grandsire of Lexington, the most brilliantly successful American stallion of the nineteenth century. Clearly the environment of Virginia was one in which Diomed himself was able to thrive, and in which the qualities of his progeny were able to express themselves fully and provide proper criteria for selection.

*Conclusions*

The lessons which may be derived from the study of genetics for the understanding of the way heredity works specifically in the thoroughbred may be summarised as follows:

1. The best stallion, or mare, is one who possesses in double strength the largest number of genes making for excellence in the racehorse. Any stallion or mare who possessed all the important genes in double strength would get offspring that were all alike and all excellent, and any differences between them would be the result of differences in environment. A stallion or mare that possesses a large number of important genes in double strength and transmits them to all its progeny, thus breeding animals with a high degree of uniformity, is said to be prepotent.

2. The fact that the offspring of even the greatest stallions and mares have shown variations too remarkable to be accounted for by differences of environment proves that no stallion or mare has possessed, all or nearly all, the important genes in double strength.

3. The only way of discovering whether thoroughbreds of either sex are prepotent, or possess a reasonable number of important genes in double strength, is by mating them and seeing how the offspring turn out. The progeny test alone reveals the breeding value of an individual by showing whether he or she breeds pure or impure for vital characters.

4. A horse possessing an excellent genotype may fail to show

good racing form owing to a variety of environmental causes, of which accident and illness are the most obvious. A horse of this kind would be more valuable for stud than a horse with an inferior genotype who showed good form on account of a more favourable environment. Only the evidence provided by the progeny can differentiate between these two types of horse.

5. Owing to the variability of the sex cells produced by most stallions and mares, the result of the vast majority of matings is largely unpredictable. Accordingly it is no mere figure of speech to describe thoroughbred breeding as something of a lottery, and even own brothers and sisters may show marked differences. Where own brothers and sisters resemble each other closely in looks of performance, one of the parents is usually prepotent. For example the brothers Dante and Sayajirao, winners of the Derby and St. Leger respectively, were by the prepotent sire Nearco, and the brothers Persimmon and Diamond Jubilee, both multiple Classic winners, were by the prepotent sire St. Simon.

6. It is possible for offspring to show characters which were not shown by either of the parents, but were present in the parents in the form of recessive genes. These characters are inherited according to Mendelian principles. For example, a bay stallion mated to a bay mare may produce chestnut offspring if neither has the gene for bay in double strength.

7. After making due allowance for paragraphs 3, 4 and 5, it may be stated that a superior racehorse must possess many good genes, even if only in single strength, and some of his sex cells are likely to contain strong enough concentrations of the right genes for him to get a proportion of first-rate offspring. It would be surprising if any truly great racehorse—a Flying Childers, Eclipse, St. Simon, The Tetrarch, Nearco, Ribot, Sea Bird II, Sceptre, Pretty Polly or Sun Chariot—did not get some satisfactory offspring provided that he or she was fertile. This indicates the value of the racecourse test for selecting breeding stock, though the racecourse test needs to be checked against the progeny test in due course.

8. The phenomenon of crossing-over means that a horse or mare may transmit chromosomes of different composition, and therefore different gene combinations, to those received from its parents. Mutation may have similar effects.

9. The fact that a thoroughbred receives half of its complement of chromosomes from each of its parents indicates that so-

called 'blood-lines', meaning the direct male and female lines of descent, are of strictly limited importance. Any particular gene is just as likely to have traced a zig-zag course through the tabulated pedigree in order to arrive at the animal in question as to have passed through the direct male or female line.

10. The contribution to a horse's genotype of a single ancestor beyond the fourth generation must be small in most cases. On the other hand the contribution of individual ancestors does not correspond to percentages which can be calculated mathematically, since the proportions are liable to be upset by such factors as segregation, dominance, gene linkage, crossing-over and mutation. A group of linked dominant genes may persist through a number of generations, though not necessarily in the direct male or female line.

11. It is not the individual as such, but his genotype, that counts in a pedigree. Any attempt to repeat the excellence of a distant ancestor by including his name in mating plans is doomed to failure unless the characters sought have been found in some of the intervening generations. If the characters sought have not been shown in intervening generations they can only be repeated if they are due to a recessive gene or genes, in which case the odds against success are too long to justify the attempt.

# 4

## *Breeding Theories*

AS RACING DEVELOPED AND BECAME part sport, part business, as the financial rewards for success and penalties for failure increased, so did the competition among breeders become ever more intense. The foundation of commercial studs with yearling sales as their principal outlet was symptomatic of this process and it was a commercial breeder, William Blenkiron, who became in 1866 one of the earliest race sponsors by giving £1,000 for a two-year-old race named after his Middle Park Stud. The prize, it is presumed, was an allowable expense to advertise the stud, and exemplified the new spirit in racing.

In this intensely competitive bloodstock world breeders were constantly in search of any means, from a magic formula to an infallible system, to produce better horses. An atmosphere was created in which breeding theories proliferated. Many of these have obtained a permanent place in the intellectual background of thoroughbred breeding and have had their devoted adherents from generation to generation of breeders. For this reason a brief analysis of the most important theories that have influenced the minds of breeders can hardly fail to illuminate the whole subject.

### *Bruce Lowe*

The theorist whose work was most widely known and had the greatest influence was the Australian Bruce Lowe. He was investigating the problems of thoroughbred breeding during the last decade of the nineteenth century, and the fruit of his researches, *Breeding Race Horses by the Figure System*, was expanded and interpreted by the journalist William Allison.

D

Lowe's method was to trace the pedigrees of all the mares included in the General Stud Book of his own time back through the female line to their original foundation mares in the earlier volumes. His findings, amended by those of Allison, were that there were about fifty of these original mares with living descendants in the direct female line. The descendants of each of the original mares comprised a 'family', and the families were classified according to the number of winners of the Derby, Oaks and St. Leger that they had produced. The family which had done best in this respect, that derived from Tregonwell's Natural Barb Mare, was called the Number 1 family. The Number 2 family, that of Burton's Barb Mare, had produced the next greatest number of winners of the qualifying races, and so one. The first five families were called the 'Running Families', and were considered to be superior to the rest in racing merit. Another group comprised those which were supposed to be particularly strong in the sire element, and these were called the 'Sire Families'. They were the Numbers 3, 8, 11, 12 and 14 Families, so that the Number 3 Family, and the Number 3 Family only, was both a Running and a Sire Family.

From these basic classifications many complex deductions about the way to breed a good horse and a good stallion were made. Lowe believed that superior qualities were the property of the Running and Sire Families in perpetuity. Any outstanding individual produced by one of the so-called 'Outside Families' should be mated with animals whose pedigrees were rich in running blood, and any horse must have a pedigree rich in sire strains in order to become a successful stallion.

The value of the work of Bruce Lowe lay in his painstaking research which revealed a great deal of detail about the way the thoroughbred had developed. He increased our store of knowledge of the subject immensely. Later research may have exposed some inaccuracies and some duplications among families which he regarded as separate, but on the whole his historical method was perfectly valid. It is the breeding theories which were built upon the foundations of his research that are unacceptable. The science of genetics has shown that qualities are not the property of blood-lines, whether male or female, nor can 'families' as such possess the faculty for producing prepotent sires in one generation after another. Prepotency is a matter of genotype and possessing large numbers of important genes in double strength. Moreover, the composition of a good genotype, containing many important genes in double strength, is dependent upon the two parents in each generation. If beauty is skin-deep, then it is equally true that prepotency is one generation deep.

Breeding racehorses by the figure system is nonsense, but there has been little more sense in most of the criticism that has been directed against the system. The general line of criticism has been that the running and the outside families have produced roughly the same proportion of good and bad horses according to their numbers, and that the superior record of the running families in the production of Classic winners has been due solely to the fact that they are much larger. Criticism of this kind makes the same basic, and fallacious, assumption as the Bruce Lowe theorists, which is that qualities can be the property of blood-lines. It is as irrelevant as criticising the diet of a grossly overweight man on the grounds that the vast number of éclairs that he consumes are filled with synthetic and not fresh cream. The shot simply is not on the target. The truth is that the overweight man ought not to be eating éclairs at all, and the telling criticism of the Bruce Lowe theories is that belief in the transmission in perpetuity of qualities through families is untenable.

These critics defeat their own object, because they beg the question why the running families are larger than the outside families. If anyone accepts the 'family idea' at all, he must accept the probability that the running families are large not merely because the members of them are specially prolific, but because their members have shown such admirable qualities that breeders have made it a matter of policy to preserve and increase them. To take this argument a stage further, the mechanism of inheritance as revealed by the science of genetics ensures that any large group of interrelated thoroughbreds is bound to include a proportion of useless individuals.

For most practical purposes the Bruce Lowe system has long been discredited. Yet even now advocates of inbreeding to the running families are to be found—advice which would not necessarily imply so much as a remote relationship in the case of the Number 1 Family, which has become so vast that Bobinsky and Zamoyski, in their monumental 'Family Tables of Racehorses', were constrained to subdivide it into no less than twenty-two different branches. Others see in the Bruce Lowe figures a safeguard against the arrangement of matings whose chances of producing satisfactory offspring are poor. According to this supposition, a breeder should think again on discovering that a stallion or mare he had intended to use belonged to one of the smallest or least distinguished of the outside families. This method of elimination could be a double-edged weapon in the hands of the breeder, causing him to reject for no logical reason an individual or a strain that was on the upgrade.

On the whole the breeders of the present day ignore the Bruce Lowe family numbers and concentrate rather on families derived from outstanding mares who have lived within the last fifty or sixty years. These families are usually called after the famous mares in which they originated, Pretty Polly and Sceptre, for example. On the other hand the Bruce Lowe family numbers are often quoted in private stud books, stallion registers and various publications. The convention is that the running family numbers are printed in italics and the sire families in bold type.

*Friedrich Becker*

The German Friedrich Becker was one of the fiercest critics of Bruce Lowe. In his book *The Breed of the Racehorse*, published in 1935, he expressed the opinion that it was absurd to argue that certain families were strong in the sire element and that it was possible to build up successful sire lines by inbreeding to those families. 'How is it', he asked, 'that of the large number of male members of the Sire Families—or of any other family for that matter—an infinitesimally smaller proportion have turned out successful sires than the female members as producers in their own rights?'

Becker believed in the almost omnipotent influence of the maternal element. Sire lines could not keep going under their own power. They came and went, but the great female lines kept going for ever. The few sire lines that had persisted had done so only as a result of repeated infusions of the best female influences. The superior vitality and productiveness of the maternal lines were attested by the fact that only 3 sire lines, but nearly 50 female lines, had survived to his own times. He estimated that of about 100 branches of the sire lines that had existed at various periods in their history only 9 had survived, but of the original 50 maternal families 300 branches were in existence and still ramifying.

For illustrating his conception of the impermanence of sire lines he placed great reliance upon the record of the St. Simon line, which had declined so rapidly from its eminent position at the turn of the century that it was on the point of extinction in England and was sustained only by the Rabelais branch in France. Yet this line which Becker was ready to write off has produced the greatest racehorse, Ribot, to have appeared anywhere since the Second World War. Ribot also was a Classic sire of exceptional merit. In addition the St. Simon line has produced the brilliant racehorses and stallions Prince Chevalier, Charlottesville, Princequillo, Round Table, Sicambre and Prince Taj. Alone among these, Ribot sprang from the Rabelais branch of the line. Thus the argument fell down even within Becker's own terms of reference.

As for the contrast between the large number of surviving maternal lines and the small number of surviving male lines, this is explained simply by the fact that a mare can have only one foal—twins seldom survive—a year but a stallion may get forty or more foals. For this reason brood mares have to be drawn from the widest possible field, but all but the best stallions can be eliminated. It is unnecessary to assume any superior qualities in the maternal lines as such.

There were grave internal weaknesses in Becker's arguments. Even if they had been watertight they would have been invalidated by his obsession with direct male and female lines of descent continued in perpetuity. The factor which decides the success or failure of any mating is not the male and female lines involved, but the genotypes of the two parents.

One other argument of Becker's requires attention. He surmised that during its long pre-natal union with the mare the foal was subject to continuous maternal influences which were capable of counteracting the effects of the genes received from the sire. Here again he was wide of the mark. The pre-natal union with the mare certainly is an important environmental factor in the development of the foal, but the foal's genotype is decided unalterably at the formation of the first body cell.

Becker was untroubled by self-doubt. 'The evidence and statistical matter which I have put forth', he wrote, 'is of an entirely different texture from the one usually advanced for supporting some theory. I am not theorising. I am exposing facts which permit of no turning and twisting—not even of any counter-arguments.' Firmly planted on this rostrum of self-assurance, he proceeded to talk mainly through his hat.

*Galton's Law of Ancestral Contribution*

Sir Francis Galton based his conclusions on statistics drawn from the study of a hundred and fifty families of human beings, covering three generations and having regard to height, eye colour, artistic faculty and health. He also analysed the inheritance of coat colour in Basset hounds. His conclusions may be expressed as follows: 'The two parents between them contribute on the average half of each inherited character, each contributing a quarter. The four grandparents contribute one-fourth altogether, and one-sixteenth each. The eight great-grandparents contribute one-eighth altogether, or one-sixty-fourth each—and so on.'

It is quite clear that this so-called 'law' is untenable as an expression of scientific truth. We have seen that the mechanism of inheritance is not controlled according to rigid mathematical proportions, either absolutely or on an average.

On the other hand it is always possible that what will not pass muster as scientific truth may have some useful practical application. Galton's Law has been applied to the case of the thoroughbred mainly as a method of assessing potential stamina in a racehorse or in the putative offspring of a projected mating. The data used are the longest distances over which the various ancestors concerned were able to win. Let us assume, for example, that the two parents of an animal under examination were X and Y, and that the longest distance over which X won was 12 furlongs and the longest distance over which Y won was 6 furlongs. This gives an average distance for the first generation of 9 furlongs, but this must be halved to produce the figure of potential stamina contributed to the animal by this generation. The contribution of the second, third and fourth generations is worked out in the same way in accordance with the proportions set out in Galton's Law, and the contributions for these four generations are then added together to produce what is known as the 'stamina index', or the distance in furlongs which it may be expected to stay. The contributions beyond the fourth generation are regarded as too small to be worth bothering about, though something ought to be allowed for them and this may amount to approximately half a furlong in a middle-distance pedigree.

The stamina index produced by the application of Galton's law may be reasonably accurate in a majority of cases. It needs to be in order to justify the considerable labour involved. There is a disquieting possibility that an equally accurate assessment of potential stamina may be made by more empirical, less complicated, methods.

It is not necessary to look very far to find cases in which the stamina index has been misleading. Tudor Minstrel won the 2,000 Guineas in brilliant style by 8 lengths. He appeared to be a class above any of the other three-year-olds of 1947. His stamina index was about $11\frac{1}{2}$ furlongs for the first four generations and, allowing half a furlong for the rest of his pedigree, which seemed fair enough, he could be counted on to stay the $1\frac{1}{2}$ miles of the Derby. In the event he started at 7 to 4 on, but palpably failed to stay and finished only fourth behind Pearl Diver. Afterwards he proved unable to stay even $1\frac{1}{4}$ miles and was beaten by Migoli in the Eclipse Stakes.

In contrast with Tudor Minstrel, St. Paddy had a stamina index of about $9\frac{3}{4}$ furlongs. Here again the figures proved grossly misleading, because St. Paddy stayed well enough to win not only the Derby but the St. Leger too.

The method of calculating the stamina index is not based on sound premises, but investigation of the examples of Tudor Minstrel and St. Paddy reveals how it is liable to go astray even as a rule-of-thumb guide. Tudor Minstrel had, on paper, all the ingredients necessary to produce a mile-and-a-half performer, but he lacked the right temperament. He was a thrustful, impatient type of colt, and refused to settle down and conserve energy in the manner that is essential to stay the distance in the top class. The question of temperament falsified the figures in the opposite way in the case of St. Paddy. His sire Aureole did not win over a longer distance than $1\frac{1}{2}$ miles because he was headstrong, though otherwise he could have been expected to stay $1\frac{3}{4}$ or 2 miles. St. Paddy's dam Edie Kelly gained her only victory over 5 furlongs, though her sire was the Derby winner Bois Roussel. As a result St. Paddy obtained an unrealistically low contribution, from the point of view of the stamina index, from the first generation of his pedigree.

The doubts raised by Tudor Minstrel and St. Paddy about the validity of the theory of the 'stamina index' suggest a further question: 'Can the staying power of a horse be properly expressed by the longest distance over which it has won?' The answer must be in the negative as far as large numbers of horses are concerned. Many good horses with limited stamina manage to win over 1½ miles or further against opposition too weak to provide a real test, and others are capable of staying much longer distances than those over which they are actually successful. Alcide would probably have won three out of every four Ascot Gold Cups, but was beaten by Wallaby II in that race over 2½ miles, and is in the records as having gained his longest victory in the St. Leger over 1 mile, 6 furlongs and 132 yards.

Stamina cannot be measured like petrol from the pump. It is not an absolute and unalterable part of a horse's make-up. Like wine, and unlike petrol, it may mature with age. If Honeyway had gone to stud, as he might have done in ordinary circumstances, at the end of his four-year-old season, he would have done so as a horse who had not won over a longer distance than 6 furlongs. When he was syndicated for stud purposes at the end of his five-year-old season he had won the Champion Stakes over 1¼ miles, but the prospectus stated that 'his testicles have not descended to the normal extent for his age'. He proved incapable then of carrying out his stud duties, with the result that he was brought back into training as a six-year-old and won over 1½ miles. By that time he had matured sexually and was able to go to stud and achieve normal fertility. From the point of view of the stamina index he represents 12 furlongs in the pedigrees of his descendants, but if he had attained early sexual maturity and gone to stud after his four-year-old season he would have represented only half as many furlongs.

The conclusion must be that Galton's Law has little of value to offer to the breeder or the student of the thoroughbred. On the contrary, what it does offer is an immensely laborious method of arriving at an estimate of stamina that is as likely as not to be false.

### *The Vuillier Dosage System*

The French Colonel Vuillier carried out intensive research during the first quarter of the twentieth century into the pedigrees of about 650 of the best racehorses, including the Classic winners, from the early days of the Turf down to his own time. He observed that certain horses appeared far more often in these pedigrees than others, and from his observations drew a number of conclusions about the breeding of a 'standard' good horse.

Vuillier's method of analysis was to set out the pedigree of the horse under examination down to the twelfth generation. There are 4,096 individuals in the twelfth generation, and to each of these Vuillier allowed a value of one. He allowed a value of two to each of the 2,048 ancestors in the eleventh generation, and so on until in the first generation each of the two parents was allowed a value of 2,048. By applying this method to all his 650 pedigrees, obtaining an aggregate value for each ancestor and averaging, he was able to calculate not only the individuals that should appear in the pedigree of a standard good horse but also their proportions, or 'dosages'. Transferred to the realm of practical breeding, his theory was that whenever a stallion or mare was lacking or deficient in any of the constituents of a standard good pedigree, a mate should be selected who would help to correct the dosages in the expected offspring.

In his last book, *Les Croisements Rationnels dans la Race Pure*, published in 1928, Vuillier, who died in 1931, based his normal dosages for breeding a good horse on fifteen horses and one mare divided into three series. The Birdcatcher series consisted of the horses Birdcatcher, Touchstone, Voltaire, Pantaloon, Melbourne, Bay Middleton and Gladiator, and the mare Pocahontas, foaled between 1824 and 1837. The Stockwell series consisted of Stockwell, foaled in 1849, and Newminster, foaled in 1848. The St. Simon series consisted of St. Simon, Galopin, Isonomy, Hampton, Hermit and Bend Or, foaled between 1864 and 1881. The dosage, or the target figure which should be aimed at in the produce of every planned mating, of each of these horses was added.

It was accepted that the series and the dosages laid down could not be permanent or immutable. As the earlier horses receded further into the past their dosages were bound to alter until finally they disappeared beyond the confines of the twelfth generation. They must be replaced by new individuals and new series. This consideration prompts the first criticism of the system, which is that the dosages are in a continuous stage of flux and that the breeder trying to apply them to his mating plans at a given point in time is like a blind man groping for an object moving past him on a conveyor-belt. By the time that the breeder's plans materialise the dosages of the ideal pedigree may have changed. The breeder then, to vary the analogy, is like a hapless politician who does his best to toe the party line, only to find that the party line has changed. In fact Mme Vuillier, the Colonel's widow, worked out new series to replace the Birdcatcher, etc., series for mating the Aga Khan's mares.

This is not the only fault that can be found with the dosage system. Some critics have pointed out that the system incorporates an elementary mathematical error in that it is assumed that each of the twelve generations contributes the total value of 4,096 points. If the two parents contribute the whole inherited make-up, what is left for the other eleven generations to contribute? In this respect the Vuillier basis for assessing ancestral contribution is in conflict with Galton's Law, which states that the two parents between them contribute only half. On the other hand it would be a quibble to claim that this inherent contradiction in Vuillier's method invalidated the whole concept of dosages if these were a fair reflection of the way that heredity works. The telling criticism of the dosage system is that it is scientifically unsound.

The clue that should lead to the exposure of the shortcomings of the Vuillier dosage system lies in the enormous differences that often exist between own brothers and own sisters, between a Triple Crown-winning Gainsborough and a useless Baydrop. The explanation of these differences in segregation, which makes it possible for parents to transmit a great variety of gene combinations to their offspring. Any system of assessing ancestral contribution which depends on rigid mathematical proportions obtained from a number of generations is untenable.

A number of breeders, notably the Aga Khan while he was building up his stud in the period between the two world wars, have used the Vuillier dosage system in planning their matings. The Aga Khan was one of the most successful breeders in the history of the thoroughbred, but his success does not justify faith in the system. He was able to use the best material, expertly chosen without regard to expense. At least during the later years of his life, the dosages were only one of the factors considered when the matings of his mares were planned, and the physical characteristics of the horses concerned were considered closely. In all the circumstances, the question of the truth or otherwise of Vuillier's theories was probably irrelevant.

In an attempt to assess the value of the dosage system, H. E. Keylock examined the complete stud records of eight well-known mares—Scapa Flow, Jennie Deans, Selene, Mumtaz Mahal, Plymstock, Friar's Daughter, Lady Wembley and Resplendent—and published his findings in his book *The Mating of Thoroughbred Horses* in 1942. His method was to collate the ability shown by the various offspring with the degree to which the matings that had produced them had followed the requirements of the dosages. His verdict was: 'Nothing can be found in the results which supports Vuillier's dosage figures.'

Few modern breeders have much faith in dosages. Those who still pay some attention to the system mostly admit that its basis may be unsound, but claim that it provides a useful means of ensuring that nothing important is left out when a mating is planned. In other words, the breeder is like a cook checking a list of ingredients before beginning to bake a cake. This conception of the function of the breeder, to blend ingredients, embodies the same fundamental error as the Vuillier system itself.

No intimation of his own fallibility bothered Vuillier. He stated: 'I can say to those who may be tempted to deny the law of dosage: "It is so—and your denial will not alter it, for it is in nobody's power to prove that truth is error." '

In recent times, Franco Varola has undertaken a revision of Vuillier's work in his *New Dosages of the Thoroughbred.* Varola acknowledged the force of criticism of Vuillier's theories when he wrote in *The British Racehorse*:

'I am deeply aware of Mr Peter Willett's warning . . . that to breed on dosages is like to try to reach an object sliding past on a conveyor-belt, and it was to avoid this danger to the utmost that I tied my dosage method to a permanent line, and I further propose to bring it up to date at least every five years.'

Varola has disclaimed any intention to present the dosage method as a magic formula for producing top class racehorses. His claim is much more modest, and is that it is an aid to practical breeding. He has added a new dimension of typology to Vuillier's ideas. His central theme is that the Classic stallions form an élite within the thoroughbred population, but this élite draws strength from continuous intercourse with complementary throughbred types designated 'Brilliant', 'Intermediate', 'Stout' and 'Professional'. The classification of stallions under these various heads is bound to be arbitrary and cannot be based on absolutely valid assessments of their merits.

The principal claim for Varola's dosage method is that it is a control system for large-scale breeding operations, and that it can help to ensure that balance is maintained and disproportions are avoided in pedigrees intended to produce racehorses of superior ability.

*J. B. Robertson*

James Bell Robertson died in 1940 at the age of eighty. He was frequently granted the courtesy title of 'Professor', and was a vet who gave up practice to devote his energies wholly to journalism. He was a regular contributor to the *Sporting Chronicle*, the *Sunday Times* and *The Bloodstock Breeders Review*, and was recognised as the leading authority of his time on the breeding of the racehorse. He was an indefatigable researcher, and his articles were models of accuracy and clear exposition.

The vital importance of Robertson's work lay in relating the discoveries of the science of genetics to the special case of the thoroughbred. Part of his contribution to *Flat Racing* in the 'Lonsdale Library' was entitled significantly: 'The Principles of Heredity applied to the Racehorse'. This was written not long before he died, and may be regarded as a summary of the arguments he had been developing during the previous thirty years. Before Robertson, there had been a free-for-all atmosphere in which the wildest theories about breeding winners had a chance to flourish; after Robertson, there was little justification for belief in the Bruce Lowe system or any other system which was not firmly based on the principles of genetics.

Robertson denied specifically the validity of the Bruce Lowe theories. He wrote in *The Bloodstock Breeders Review*:* '. . . the fact that none of the ordinary chromosomes is chained, as it were, to either the male or the female lines, renders absolutely nugatory the Bruce Lowe doctrine of perpetual anchorage in the direct female line of those characters and qualities which make or mar a racehorse.'

Unfortunately for his scientific reputation, he was unable to adhere steadfastly and logically to the implications of his own knowledge. He realised that none of the ordinary chromosomes was chained to either the male or the female line; but he was mesmerised by his awareness of the process of sex inheritance. He wrote in the same volume of *The Bloodstock Breeders Review*: 'In respect of the Y chromosome, there is continuity in descent in the male line chain. On the other hand in the female, mother to daughter line, the chain is not unbroken either through the X or ordinary chromosomes.' He pointed out that, for example, Sceptre carried a paternal X derived from Persimmon's dam Perdita II, but her maternal X may have been derived

* Volume XXVI, 1937.

from either Bend Or's dam Rouge Rose or from Ornament's dam Lily Agnes. Sceptre was by Persimmon (by St. Simon out of Perdita II) out of Ornament, by Bend Or (by Doncaster out of Rouge Rose) out of Lily Agnes. His conclusion was that, although the X and Y chromosomes contained few, if any, of the genes which determine characters unconnected with sex, the male line is a safer guide than the female line. Clearly he had a prejudice in favour of the male line in pedigrees, and was not ruthless enough in confronting his prejudice with his scientific knowledge.

This aberration or, perhaps, failure of the will, was a recurring theme in his writing. His summary of the considerations which should determine mating plans was in most respects unexceptionable. He laid down the rule that it is essential to make a close study of the characters, good and bad, of the proposed parents; and added that it is useless to expect benefits from a distant ancestor in a pedigree unless some of the characters of that ancestor have been shown in intervening generations. The rule is sound. He was moving on to more controversial ground when he went on to say that first-hand knowledge of at least the male ancestors in the last fifty or sixty years is needful. The number of years is irrelevant, because it is generations, not years, that count in heredity. Yet this mistake is trivial in comparison with the fundamental error, here implied, of assuming that the sire is of greater importance than the dam in any given mating. The truth is that sire and dam are of equal importance, granted the qualifications that are described in a later chapter.

Thus Robertson fell short of perfection as an interpreter of genetics in terms of the thoroughbred; at the same time, all modern students of the racehorse owe him an enormous debt. Without his work, thoroughbred breeding would still be, scientifically speaking, in the dark ages.

# 5

# *St. Simon*

A DISTINGUISHED FOREIGN VISITOR, on entering St. Simon's box while the horse was at stud at Welbeck, raised his hat in homage to the greatest thoroughbred that had ever lived. The claim of St. Simon to that title may, of course, be disputed. Comparisons between horses of different eras are bound to be matters of opinion to some extent. Yet the excellence of a thoroughbred must be judged neither by racing performance nor by stud achievement alone, but by a combination of both these factors and, by this double standard, St. Simon must rank very high indeed. An analysis of his racing and stud careers may help to establish his claim; and if his claim appears to be valid, then it follows that his career as a whole, his pedigree, his own qualities, and the qualities of his progeny and descendants, should yield a rich harvest of information about the working of heredity in the special case of the racehorse.

St. Simon was a bay horse, foaled in 1881, by Galopin out of St. Angela. He was bred by Prince Batthyany, a Hungarian by birth who made his home in England and became a successful owner and breeder. He won the Derby with Galopin in 1875, though he did not breed that horse but bought him as a yearling. He was an old man and suffering from a bad heart when St. Simon was foaled, and died of a heart attack, presumably brought on by excitement, as he was walking towards the Jockey Club Luncheon Room on the Rowley Mile course at Newmarket half an hour before Galliard, owned by Lord Falmouth but sired by his beloved Galopin, won the 2,000 Guineas in 1883. St. Simon, then two years old, was bought by the Duke of Portland for 1,600 guineas at the dispersal sale of the Prince's bloodstock in July the same year.

In accordance with the rules in force at the time, all St. Simon's engagements became void on the death of his nominator. Thus there was no question of his running in any of the Classic races, which had closed when he was a yearling, though one lady to whom he was shown late in his stud career at Welbeck remarked that she distinctly remembered him winning the Derby twice! In fact he was not even entered in the Derby, or the St. Leger, his only Classic engagement being in the 2,000 Guineas. The reason why he was not entered fully in the Classic races is not known for sure. One theory was that Batthyany took a great dislike to the Derby course after the victory of Galopin. A more plausible explanation is that Batthyany thought that the colt was unlikely to attain the required standard. St. Angela had shown only third-class racing form and had produced only one foal of note before she bred St. Simon at the age of sixteen; that was St. Simon's full-sister Angelica, who was destined to achieve fame much later as the dam of Orme, a winner of the Eclipse Stakes twice. There was no strong reason to anticipate the excellence of St. Simon by entering him for the Derby.

St. Simon had not run when Portland bought him. He was hurriedly entered in such suitable races as had not already closed, and began his racing career by winning the Halnaker Stakes and a maiden race at Goodwood, both with the greatest ease. He ran next in the Devonshire Nursery at Derby, in which he carried top weight and was described as 'winning in a canter by two lengths'. He then went on to Doncaster for the Prince of Wales's Nursery, in which he carried 9 stone and his twenty opponents carried weights from 7 stone 11 pounds down to 5 stone 12 pounds. Archer had a fixed prejudice against winning by big margins, but on that occasion he could not restrain St. Simon, who passed the post eight lengths in front of his nearest rival, who happened to be Portland's second string Iambic, carrying 6 stone 7 pounds. It may be remarked parenthetically that this same Iambic met Asil, the best Arab of his day, in a race over the last three miles of the Beacon course at the Newmarket Second Spring Meeting two years later. Giving Asil 4 stone and 7 pounds, Iambic won by twenty lengths and thus provided an eye-opening demonstration of the extent to which the thoroughbred was superior in speed to the Arab stock from which he had evolved. The superiority of St. Simon to Asil is beyond accurate computation, but must have been 10 stone at a conservative estimate.

St. Simon's last race as a two-year-old was a match against the Duke of Westminster's Duke of Richmond at the Newmarket Houghton Meeting. On his only previous appearance Duke of Richmond, formerly called Bushey, had gained an easy victory in the Richmond Stakes at Goodwood and was regarded as just about the best of his age. Tom Cannon had orders to jump off and make the running for the whole of the six furlongs on Duke of Richmond. Horse and jockey both did their best to carry out these instructions, but after going a quarter of a mile St. Simon was fifty yards in front of them. Archer then dropped his hands on St. Simon and allowed his opponent to come much closer, so that St. Simon won in a canter by three-quarters of a length. The next year Duke of Richmond was second in the Royal Hunt Cup carrying 8 stone, in the Wokingham Stakes carrying 8 stone 11 pounds, and in the Stewards Cup carrying 8 stone 10 pounds, so St. Simon had toyed with a horse who was at least a high-class handicapper.

During the winter the advice of Mat Dawson, Portland's very experienced trainer, was that St. Simon should have not too strenuous a second season and should then make one of the finest stallions the world had ever seen. Accordingly the campaign was planned with the Ascot Gold Cup as the only important objective, but St. Simon began his three-year-old season by taking part in one of the most unorthodox events in the annals of the Turf. It was a so-called 'Trial Match' over the last 1½ miles of the Cesarewitch course at the Newmarket Second Spring Meeting; but it was not a match, because four horses took part, and it was not a trial, because the result was reported in the Calendar. The purpose of the race was to try St. Simon with Tristan, a top-class six-year-old who had won the Ascot Gold Cup and the Hardwicke Stakes the previous year, and the Hardwicke Stakes and the Gold Vase in 1882 also. Credo for Tristan, and Iambic for St. Simon, were put in the Trial Match as pacemakers. When the pace was turned on in earnest the two pacemakers ceased to count, and in the final stages St. Simon came right away to beat Tristan easily by six lengths.

St. Simon was so impressive in the Trial Match that he was given a walk-over in his first regular race as a three-year-old, the Epsom Gold Cup. His next race was the Ascot Gold Cup, in which his principal opponent was again Tristan whom he met, as in the Trial Match, at weight for age. The result, though perhaps not the manner of it, was a foregone conclusion. Wood rode St. Simon because Archer could not do the weight, and was told not to win by more than two lengths. Through most of the race Wood kept St. Simon in check well enough, but St. Simon took matters into his own hands when Wood let out a reef after turning into the straight, and sprinted right away to win by twenty lengths. Some people were inclined to conclude that Tristan must have deteriorated, but Tristan scotched that idea by turning out again the next day to win the Hardwicke Stakes for the third time.

St. Simon had only two more races, and winning them was nothing but a formality. They were a Gold Cup over a mile at Gosforth Park, and the Goodwood Cup, in which he beat his two opponents in a canter by twenty lengths.

In terms of cold figures the racing career of St. Simon did not amount to a great deal. He ran in, and won, nine races, excluding the Trial Match, of a total value of £4,676 15s. To put the stakes in perspective, this was about £200 less than the Derby was worth jointly to the dead-heaters St. Gatien and Harvester in 1884. But the first of the lessons to be gleaned from a study of St. Simon may be that neither the value of stakes won, nor victory in any named race, notably the Derby, provides the true criterion by which a horse should be judged. What really counts is quality of performance, and this may involve much more complex problems of assessment.

St. Simon was not extended in any of his races, but even at their face value his performances in the Prince of Wales's Nursery, in his match with Duke of Richmond, and in his two encounters with Tristan, were outstanding. All the same, his performances alone did not tell anything like the whole story and at least equal importance must be granted to the impressions which he conveyed to those most intimately connected with him. His trainer Mat Dawson once observed: 'I have trained only one smashing good horse in my life—St. Simon'; and Dawson won the Derby with Thormanby, Kingcraft, Silvio, Melton, Ladas and Sir Visto, besides the Grand Prix with Minting. He amplified this statement on another occasion when he said: 'The extraordinary thing was that St. Simon was at good at a furlong as he was at three miles, and distance never seemed to worry him.' The trainer told Portland that he had never seen a horse so full of what he called 'electricity', and that he even moved in his box in a different manner from any other horse he had ever trained. John Huby, Portland's stud groom at

Welbeck, was expressing much the same thought when he said of St. Simon: 'I always thought that there was something superior, both in his action and contour, to anything I had ever noticed in any other horse.' Observers were equally impressed by his 'electricity', or 'innate vitality', and by his action. He was described variously as moving 'like a greyhound' and 'as if made of elastic'.

*Stud Career of St. Simon*

In accordance with Mat Dawson's recommendation, St. Simon did not run after his three-year-old season. He was rested in 1885 and took up stud duties at Dawson's Heath Stud at Newmarket the next covering season. Later he was moved to Portland's own stud at Welbeck, where he remained until, at the age of twenty-seven, he dropped dead as he was returning from exercise on 2nd April 1908. His fee for his first season was 50 guineas, but this was doubled the next year and after that was increased progressively to reach a high point of 500 guineas in 1900, at which it remained. In his first season he got 16 of the 20 mares he covered in foal; this represented a fertility percentage of 80, and during the whole of his stud career he covered 775 mares and got 554 in foal, representing a fertility percentage of 71·4.

The increases in his stud fee reveal that he was held in a high esteem by breeders which cannot be accounted for purely by his extremely satisfactory fertility. His progeny showed high-class form from the time that they first appeared on the racecourse. Of his first lot of two-year-old runners, appearing in 1889, nine won thirty-four races worth £24,286 and took him into third place in the list of sires of winners of all ages, behind his own sire Galopin and Hampton. The next year he rose to the top of the list and remained there for seven consecutive seasons. He was second in 1897, third in 1898, and fifth in 1899, but returned to the top in 1900 and 1901. He was second in 1902, third in 1904, seventh in 1905 and fifth three years later, these being the only other times he appeared in the first ten sires. Altogether he sired the winners of 571 races of £553,158.

These figures would have represented phenomenal success in any era of Turf history, but the magnitude of his achievement as a stallion may be expressed even more graphically in terms of the Classic victories gained by his progeny. Altogether ten of his sons and daughters won seventeen Classic races, as follows:

Persimmon and Diamond Jubilee won the Derby; Memoir, La Fleche, Mrs. Butterwick, Amiable and La Roche won the Oaks; Memoir, La Fleche, Persimmon and Diamond Jubilee won the St. Leger; St. Frusquin and Diamond Jubilee won the 2,000 Guineas; and Semolina, La Fleche, Amiable and Winifreda won the 1,000 Guineas. The year 1900 represented the climax of his career as a stallion because in that season his progeny carried off all the five Classic races, Diamond Jubilee winning the Triple Crown, Winifreda the 1,000 Guineas and La Roche the Oaks.

The bare facts are eloquent testimony to St. Simon's greatness, but they must be clothed with some details of his own characteristics and those of his offspring in order to complete the picture. He stood 15 hands 3¼ inches as a three-year-old, but grew to 16 hands while he was at stud. He was so perfectly proportioned that many observers took him to be smaller than he really was; he had great quality, an exceptionally concave profile, shoulders so sloped that he appeared to be short in the back, and remarkable length from hip to hock. He was higher at the croup than at the withers, an unusual characteristic which he shared with another of the outstanding horses of thoroughbred history, Eclipse; this may have given them extra leverage and helped to account for their brilliant speed. He was a grand doer and had such a magnificent constitution that he was hardly ever sick or sorry in the whole of his life.

St. Simon's 'electricity', or 'innate vitality', was expressed in irritability and free sweating. Many people thought him bad-tempered, but Huby denied this, saying that he was really a good-tempered horse, but his highly-strung temperament necessitated the most gentle and patient treatment. There was probably a vein of euphemism in Huby's statement, as the stud groom obviously hero-worshipped the horse. Visitors to his box at Welbeck sometimes found his good temper less evident, though he had his weakness and could be kept at bay by being confronted by an umbrella or a hat held up on the end of a stick.

St. Simon passed on his highly strung temperament and tendency to sweat freely to many of his progeny. He also transmitted his good looks and constitution, and much of his superb speed and stamina. His stock were known for being exceptionally clean-winded; his sons and daughters needed little work to get them fit, and most of them were wonderful movers. Portland remarked that 'the stock of Galopin and St. Simon were more nearly Robots than any I have known before or since'. Some even of the best of them were small, and Mrs. Butterwick stood only 15 hands when she won the Oaks. Theodore Cook stated, undoubtedly with justice, in his *History of the English Turf*, that St. Simon and his progeny set the pattern by which trainers were apt to judge horses for many years.

For some years after he went to stud there was a growing belief that St. Simon's fillies were better than his colts. His fillies Semolina, Memoir, La Fleche, Mrs. Butterwick and Amiable all won Classic races before a single colt by him was successful in a Classic race; but then St. Frusquin and Persimmon came along in 1896 to win all the three Classic races for which colts were eligible and consign prejudice to the refuse heap of ideas where it belonged. There have been instances of stallions whose stud records as a whole proved that their progeny of one sex were conspicuously better than their progeny of the other sex—Aureole's colts on balance were much superior to his fillies. But in most cases the merits of the progeny of the two sexes tend to even out in the long run, like the incidence of heads and tails with a tossed coin or red and black in roulette, and conclusions based on the results of the first few seasons of a stallion's stud career are usually premature and are apt to be falsified by later events.

### *St. Simon's Place in Turf History*

How may St. Simon be compared with other great racehorses and sires of thoroughbred history? Perhaps he was no greater than Flying Childers, or Eclipse, or Ormonde, or The Tetrarch, or Nearco, or Gladiateur as a racehorse, judging each by the standards of performance of his own time. Only between St. Simon and Ormonde, the unbeaten Triple Crown winner foaled in 1883, can any sort of valid comparison be made; for Fred Archer rode them both in most of their races, and had no doubt that St. Simon was the better. When achievement at stud also is taken into consideration, all vestige of doubt disappears at once, because Ormonde passed on his own wind infirmity to many of his offspring and soon became infertile. St. Simon also excelled the other five horses mentioned as a sire—even Eclipse who, as we have seen, had serious rivals for the honour of being the best sire of the late eighteenth century.

The so-called 'Emperor of Stallions' Stockwell, Hermit and

Hyperion are horses worthy of comparison with St. Simon in the category of great sires. Stockwell and Hermit each headed the list of winning sires seven times, twice less than St. Simon. Stockwell, foaled in 1849, got the Derby winners Blair Athol, Lord Lyon and Doncaster, and was responsible for the winners of seventeen Classic races altogether, the same total as St. Simon. Hermit, foaled in 1864, got the Derby winners Shotover and St. Blaise, but his sons and daughters won a total of no more than seven Classic races. Hyperion, much the most recent of the bunch, was foaled in 1930. He was the leading sire of winners six times, but got only one Derby winner, Owen Tudor, and the winners of eleven Classic races altogether. As stallions, there may not have been a great deal to choose between Stockwell and St. Simon, but in respect of racing merit the comparison must be favourable to St. Simon. Although Stockwell won the 2,000 Guineas and the St. Leger, he was unplaced in the Derby and was beaten in the Ascot Gold Cup the next year. Hermit, a blood-vessel breaker, won the Derby, but ended his racing career with an unbroken series of fourteen defeats. Hyperion won the Derby and St. Leger but failed badly in the Ascot Gold Cup as a four-year-old. None of these three had the aura of invincibility which surrounded St. Simon.

### *The Pedigree of St. Simon*

In the final analysis there is no need to insist on the absolute supremacy of St. Simon, for there can be no reasonable doubt that he is entitled to a place on the shortest of short lists not only of the greatest racehorses but of the foremost prepotent sires also—and was entitled, too, to the homage of his distinguished visitor. Thus we move on to the question: 'What lessons may the student of breeding draw from his pedigree and achievements?'

The salient facts of the pedigree of St. Simon, including those close ancestors who showed top-class racing form, may be set out quite easily. Taking the sire's side of the pedigree first, St. Simon was by Galopin, who may be placed in his proper category simply by stating that he was one of the eight 'smashers' in Dr. Platt's respected classification of the Derby winners from 1860 to 1911 in Volume I of *The Bloodstock Breeders Review*. Although the prejudice of many breeders retarded his success at stud, Galopin eventually forced his way to the front by sheer merit and got, besides St. Simon, the 2,000 Guineas winners

Galliard and Disraeli, the 1,000 Guineas winner Galeottia and Donovan, who won the Derby and St. Leger.

Galopin had as his sire the 2,000 Guineas and St. Leger winner Vedette, and as his two grandsires Voltigeur and The Flying Dutchman, who both won the Derby and St. Leger. One of the romantic twists that have occurred at frequent intervals in Turf history is that Voltigeur and The Flying Dutchman, deadly rivals in two of the most celebrated of matches, should have been concerned jointly and intimately in the production of horses as eminent as Galopin and St. Simon. The Flying Dutchman gained his Classic victories in 1849, and Voltigeur his a year later. They met for the first time in the Doncaster Cup two days after Voltigeur had won the St. Leger, the two of them being the only runners. Voltigeur won by half a length, but few observers accepted the result as conclusive and a second meeting was arranged at York the following May. The handicapping was entrusted to Admiral Rous, who set The Flying Dutchman to give Voltigeur 8½ pounds, which was exactly the amount recommended in his weight-for-age scale for a five-year-old to give a four-year-old over two miles at that stage of the season. This time The Flying Dutchman prevailed by a distance described as a short length to prove that they were two very good horses and practically equal in merit.

Galopin's dam Flying Duchess and St. Simon's dam St. Angela and granddam Adeline were all greatly inferior to these excellent animals as performers on the racecourse. Flying Duchess was the only one of the three to win more than one race; she was successful in a nursery over four furlongs at the Newmarket Second October Meeting as a two-year-old and in a selling handicap over the same distance at the First Spring Meeting the next year. St. Angela gained her only success in a £70 maiden plate over five furlongs at the Newmarket Second October Meeting as a two-year-old; she managed to dead-heat with The Sphinx in a £50 fillies sweepstakes at the next Craven Meeting, but was beaten in the run-off. The sole victory of Adeline was in the Tradesmen's Handicap over five furlongs at Yarmouth as a three-year-old. They were very moderate racehorses.

St. Angela and Adeline, on the other hand, were not properly representative of the quality of the female half of the pedigree of St. Simon. It is necessary to go back only one more generation in the direct female line to find as Adeline's dam Little Fairy, a half-sister of the Derby winner of 1840, Little Wonder. St. Angela's sire King Tom and her two grandsires Harkaway and Ion were all top-class performers. King Tom, a half-brother of Stockwell, might have won the 1854 Derby instead of finishing second to Andover but for a training setback shortly before the race. Without matching the greatness of Stockwell, he did very well at stud and was leading sire twice. His progeny included the Derby winner Kingcraft, the 1,000 Guineas winner Tomato, the Oaks winners Tormentor and Hippia, and Hannah, who joined the select band of fillies who have won three Classic races by winning the 1,000 Guineas, the Oaks and the St. Leger.

King Tom's sire Harkaway had an unconventional career. He raced only in Ireland at two and three years of age and there won most of the long-distance races available. His last two seasons in training were spent in England, where he proved himself a fine stayer by winning the Goodwood Cup twice. Ion, the maternal grandsire of St. Angela, was second in the Derby and the St. Leger in 1838, and sired the Derby winner Wild Dayrell.

The pedigree of St. Simon may be summed up by stating that the first three generations included three Derby winners, of whom one was a 'smasher' and the other two won the St. Leger also; one winner of the Guineas and the St. Leger; two Derby runners-up, of whom one was second in the St. Leger also; and one exceptionally good and hardy stayer—these accounting for all the seven sires concerned. Of the mares concerned, one was a half-sister of a Derby winner and another, King Tom's dam Pocahontas, produced not only King Tom himself but the outstanding racehorse and sire Stockwell and Rataplan, who was third in the St. Leger, fourth in the Derby and won the Ascot Gold Vase and the Doncaster Cup.

This pedigree represents so dense a concentration of racing and breeding ability that it may be accepted readily as a likely source of a brilliant horse possessing a large proportion of the most important genes in double strength. But there the process of identifying the hereditary sources of his excellence comes to an abrupt halt, because there is no means of tracing the route taken by the genes, and combinations of genes, which were responsible for his peerless speed, his constitution, and his 'electricity'. It is impossible to allot fair shares of responsibility for the end-product, St. Simon, between, say, Vedette and King Tom, or Pocahontas and Little Fairy.

Colour alone provides a means of tracing one item in the

chain of inheritance, because it is visible and controlled by a single pair of genes of which one dominates the other absolutely in the vast majority of cases, and which behave in accordance with Mendelian principles. The gene for colour is transmitted independently, or linked with only a small proportion of the total number of genes; for this reason the study of colour cannot reveal how the essential racing characters have been inherited by any particular horse, in this case St. Simon: but it can throw a beam of light on one aspect of the mechanism of inheritance which would otherwise remain wholly obscure.

Both the parents of St. Simon, his four grandparents and six of his eight great-grandparents were bay or brown. The exceptions were Vedette's dam Mrs. Ridgeway, who was described as a roan, and Harkaway, who was chestnut. Of the bays and browns, Voltigeur, The Flying Dutchman, Vedette and Galopin all bred pure for that colour; that is, they held the gene for bay or brown, which dominates chestnut, in double strength, and got none but bay or brown offspring.

St. Angela, on the other hand, was an impure bay, and three of her ten offspring recorded in the General Stud Book were chestnuts. By employing the symbols set out in Chapter Three, St. Angela may be represented as BC and Gaolpin as BB and, according to Mendelian principles, these two when mated should have produced offspring in the ratio of two pure bays to two impure bays. In other words, the odds were no better than even that St. Simon would breed pure for the single character of bay colour in spite of the fact that twelve of his fourteen nearest ancestors were bays or browns themselves.

If the odds were only even that St. Simon would have in double strength the gene controlling a single character which 85 per cent of his nearest ancestors had shown themselves, it follows that the odds must have been astronomically long against his having in double strength all the large number of genes required to make a superior racehorse, and in which several of his near ancestors were conspicuously deficient. None of St. Simon's six ancestors in the first two generations was deficient in the gene for bay colour, but three of the six—St. Angela, Flying Duchess and Adeline—were deficient in some at least of the genes required to make a good racehorse. Indeed the odds must have been heavily against his inheriting a sufficiently strong concentration of the right genes in single strength in order to be a great racehorse himself, let alone in double strength so as to become

*(9)* Sceptre. Foaled in 1899, Sceptre was one of the greatest racemares of all time and won all the Classic races with the exception of the Derby.

*(10)* Orby. The Derby winner of 1907, Orby became one of the most potent influences for speed in the modern thoroughbred

*(11)* Galopin. Foaled in 1872, Galopin was one of the best Derby winners of his era, and left an indelible mark on thoroughbred history by siring St Simon

*(12)* Persimmon. The winner of the Derby and St Leger in 1896, Persimmon was St Simon's most successful son at stud and was leading sire four times

*(13)* Pretty Polly. One of the greatest of racemares, Pretty Polly won the 1,000 Guineas, Oaks and St Leger in 1904, and became the ancestress of the Derby winners St Paddy and Psidium

*(14)* The Tetrarch. Nicknamed 'The Spotted Wonder', The Tetrarch, foaled in 1911, was one of the most brilliant horses ever bred

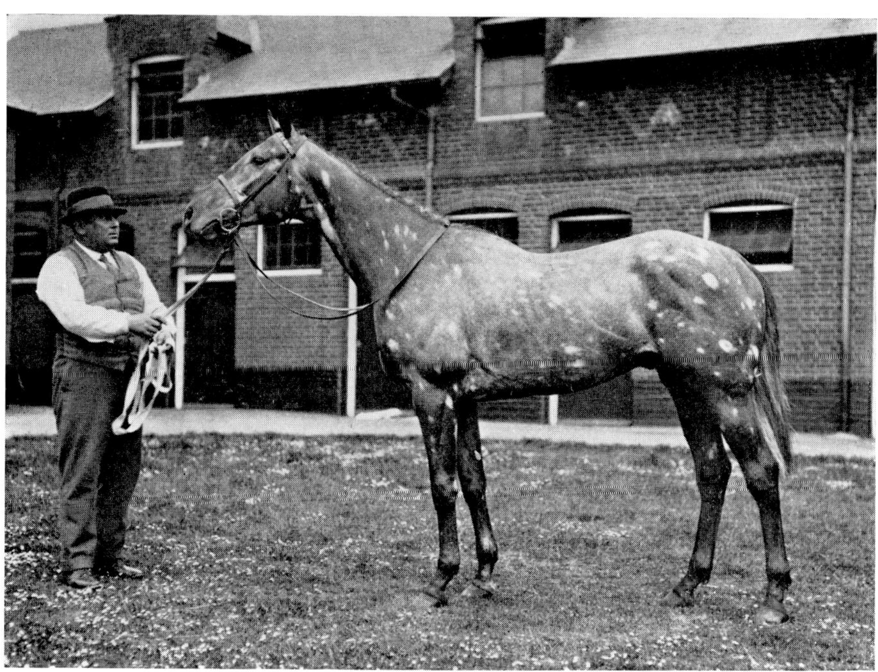

*(15)* Durbar II. The winner of the Derby in 1914, Durbar II was ineligible for the General Stud Book under the terms of the so-called 'Jersey Act'

*(16)* Blandford. Foaled in 1919, Blandford became the greatest Classic sire of the period between the two world wars

a prepotent sire. It is no wonder that a racehorse and sire worthy of mention in the same breath as St. Simon appears only once in many thoroughbred generations.

St. Simon had practically the perfect genotype. He bred pure for the only manifest and exactly measurable character, which is colour, for he got only bay or brown offspring with a solitary exception, a grey filly who was his last foal of all. That he also bred pure for many of the genes that are essential to the making of a superior racehorse may be deduced from the large numbers of his successful progeny, and the excellence shown by a high proportion of them. On the other hand, one of the fundamental problems of the breeder, to project into future generations a perfect genotype once it has been achieved, is demonstrated only too clearly by comparing St. Simon with the best of his sons. The fact is that none of St. Simon's sons attained the supremacy, either as racehorse or sire, of St. Simon himself.

The best of St. Simon's sons was Persimmon. A 'smasher' in Dr. Platt's classification, Persimmon was undoubtedly a better racehorse than his younger brother Diamond Jubilee, though Diamond Jubilee won three Classic races to his two. Diamond Jubilee made hard work of beating quite second-rate opponents, suffered several defeats and had such a bad temper that he was almost unmanageable at times. Persimmon was probably even better when he won the Ascot Gold Cup than he was when he won the Derby, but he was not invincible; St. Frusquin, whom he beat by a neck in the Derby, beat him in the Middle Park Stakes and the Princess of Wales's Stakes.

Persimmon was also the best of St. Simon's sons at stud. He was leading sire four times and got the winners of eight Classic races including Sceptre, winner of the 1,000 Guineas, 2,000 Guineas, Oaks and St. Leger. He got no colt of comparable ability, his best being Prince Palatine, who won the Ascot Gold Cup twice. Persimmon would no doubt have got more top-class horses if he had not died as a result of a fractured pelvis when he was fifteen, but in order to keep his achievement in perspective it is necessary to add that St. Simon sired the last of his Classic winners when he was fifteen, and was leading sire for the seventh time at that age.

St. Frusquin, in 1903 and 1907, and Desmond, in 1913, were the only other sons of St. Simon to head the list of winning sires. Diamond Jubilee, who was exported to the Argentine when he was nine, got into the top ten sires in England only once, being seventh in 1907. No grandson of St. Simon in the male line ever became leading sire.

It is evident that none of the sons of St. Simon, not even Persimmon, was prepotent in the same degree as St. Simon himself. Impurities began to creep in and erode the perfect genotype in the first generation of St. Simon's descendants. Persimmon, and all the less successful sons of St. Simon who stood at stud, had fewer important genes in double strength than their sire. This was due, of course, to the contribution of genes made by the mares with whom St. Simon was mated to all his progeny.

In this instance colour was again the outward sign of what was happening within the genotypes of the progeny of St. Simon. Persimmon, St. Frusquin, Desmond and Diamond Jubilee were all impure bays—that is, they all got a proportion of chestnut offspring. This impurity, the recessive chestnut gene, had been introduced into their genotypes by their dams; and we can deduce that other recessives had been introduced in the whole range of genes controlling racing performance.

St. Simon represented the climax of two hundred years of development of the thoroughbred; the difficulty was to capitalise on that achievement. The problem facing breeders trying to produce a perfect racehorse may be compared to a jigsaw puzzle, with genes representing the pieces. In the case of St. Simon the last few vital pieces had been fitted into place to complete the picture. But no breeder can halt for long to admire his handiwork. The evolution of a breed is a continuous process, and no sooner had St. Simon passed from the scene than hidden forces, uncontrolled and apparently uncontrollable, began to shake the puzzle to pieces again.

# 6

# *Inbreeding*

WHEN A JIGSAW PUZZLE IS PLACED intact in its box, and the box is shaken, the puzzle does not fall apart into all its component pieces at once. Some sections hang together at first, but if the box is shaken again and again the puzzle continues to disintegrate. If the intact puzzle is taken to represent an excellent genotype, and each shake of the box to represent a generation, it is easy to understand what is liable to happen in the case of a St. Simon and his descendants. Is there nothing the breeder can do to prevent, or at least to mitigate, this process of disintegration? The answer is that there is one method, and one method only, of projecting a particular genotype into future generations, and that is inbreeding.

### *Definition of Inbreeding*

Inbreeding may be defined as the mating of closely related individuals, as opposed to outbreeding, which means the mating of less related or unrelated individuals.

Each member of a breed has two parents, four grandparents, eight great-grandparents, sixteen great-great-grandparents, thirty-two great-great-great-grandparents, and so on. By inbreeding, the number of individuals who make up these generations may be drastically reduced. For example Mon Fils, the winner of the 2,000 Guineas in 1973, was by Sheshoon (by Precipitation) out of Now What by Premonition (by Precipitation); and had Fair Trial twice in the fourth generation of his pedigree. In consequence the 30 male places in the second, third, fourth and fifth generations of his pedigree were occupied by only 23 different individuals.

All thoroughbreds are inbred in comparison with the members of most human societies. A simple calculation will show that a modern racehorse has, in a single generation not so very far back, an aggregate of ancestors which outnumbers the entire foundation stock of the breed, and this can have come about only as a result of inbreeding. It was seen in Chapters 1 and 2 that intensive inbreeding went into the making of Flying Childers and Eclipse, and the names of Eclipse and other great sires of the same period were repeated many times in the pedigrees of all thoroughbreds of later periods. For example, the pedigree of St. Simon's grandsire Vedette contained twenty-two crosses of Eclipse and thirty-two of Herod in the first eight generations.

Inbreeding implies a closer relationship between the two animals mated than the average of the breed. In the case of the thoroughbred this implies that the ancestor common to both parents is duplicated within the first four or five generations of the offspring of the mating.

There are two methods of describing degrees of inbreeding. The first is to count the number of generations from the offspring to the common ancestor on the sire's side of the pedigree, and then to count the number of generations from the offspring to the common ancestor on the dam's side of the pedigree, and to state the two figures so obtained, separated by a multiplication sign. Thus Nearco was inbred to St. Simon 4×4, because St. Simon appeared in the fourth generation on both sides of his pedigree.

The other method is to count the generations that are free of the common ancestor. The nearest generation in which an ancestor may appear twice is that of the grandparents, and when this occurs the offspring is said to be inbred with no free generations. Thus Coronation V, a winner of the Prix de l'Arc de Triomphe and second in the Oaks, was inbred to Tourbillon with no free generations because she was by Djebel, by Tourbillon, out of Esmeralda, by Tourbillon. When the inbreeding is more remote, the number of generations intervening between the second and the common ancestor is counted on each side of the pedigree, and the two figures so obtained are added together to discover the number of free generations. Whereas Coronation V was inbred to Tourbillon with no free generations, the Champion Stakes winner Hugh Lupus was inbred to Tourbillon with one free generation because he was by Djebel, by Tourbillon, out of Sakountala, by Goya II, by Tourbillon—the count being nought on the sire's side and one on the dam's side.

Nearco was inbred to St. Simon with four free generations, this total being made up of counts of two on each side of his pedigree.

The 'free generation' method does not permit the accurate description of degrees of inbreeding even closer than that involved in the case of Coronation V. For example, the mating of a sire with his own daughter or, as in the case of Flying Childers, the mating of a stallion with his own dam (see Chapter One).

The first method of expressing degrees of inbreeding is more generally used by writers on the subject of breeding.

*The Results of Inbreeding*

The effect of inbreeding is to reduce the number of ancestors and so diminish the reservoir of genes from which the genotypes of the offspring of inbred matings are drawn. For this reason a single offspring of an inbred mating is likely to resemble the duplicated ancestor more closely than it would do if the same ancestor appeared only once in its pedigree. Moreover the various offspring of an inbred mating that is repeated several times are likely to resemble each other more closely than own brothers and sisters produced by outbred matings.

This restrictive action of inbreeding has the second important effect of increasing the number of genes held by the offspring in double strength. When any product of the inbred mating goes to stud, he will pass on the increased number of genes held in double strength to all his own offspring, not merely to some of his offspring as would be the case if he held the genes concerned in single strength.

The chief practical effect of inbreeding emerges at this point. This is the increase of uniformity within the inbred stock. Inbred animals tend to resemble their duplicated ancestors and each other more than they would if no inbreeding had taken place, and to transmit their inherited characters to the next generation with greater regularity. This revelation of the effect of inbreeding may evoke the immediate objection that the aim of the breeder of racehorses is not to achieve uniformity, but to produce a few outstanding individuals.

There are two answers to this objection. The short answer is that any breeder would surely be more satisfied if he could produce animals uniformly as good as St. Simon, Nearco and Ribot, provided that he was ahead of his rivals in that respect.

The longer answer is that some degree of uniformity is essential to enable breeders to plan their matings at all. Indeed the creation of any distinct breed of animals with the capacity of breeding true to type depends on inbreeding for the purpose of stamping the desired characters firmly in the stock. Many breeds of farm and domestic animals have been created by inbreeding of the most intensive kind. While they were developing the breed of Shorthorn cattle the Colling brothers produced the famous bull Comet by mating the bull Favourite with one of his own daughters whose mother was also his own (Favourite's) mother. To employ the phraseology of thoroughbred breeding, Comet had the same sire and maternal grandsire, and the same granddam and paternal granddam.

Enough inbreeding has been used in the development of the racehorse, as the example of Vedette indicated, to ensure that there is a good deal of stability in thoroughbred genotypes. The jigsaw puzzle can never be shaken apart into all its component pieces, even if no further close inbreeding is resorted to. Thus breeders are able to plan matings, not with certainty, but with sufficient probability as to the result to make the whole exercise worth while.

Thus there can be no argument whether inbreeding is applicable to the special case of the thoroughbred. It is merely a question whether further benefits are likely to be derived from continuing to mate individuals who are more closely related than the average of the thoroughbred population.

### *Additional Effects of Inbreeding*

It has been stated that the principal effect of inbreeding is to increase uniformity. Unfortunately this statement must be qualified because inbreeding may also have the paradoxical result of producing a certain number of individuals who are completely unlike the rest, and are worthless. It is this phenomenon which gives rise to allegations that inbreeding causes degeneracy. The truth is that inbreeding as such never causes degeneracy; it merely lays bare faults which have been hidden in the parent stock in the form of recessives. To explain this phenomenon it is necessary to abandon the metaphor of the jigsaw puzzle and instead to compare a genotype to a raincoat consisting of two layers of waterproof material. The two layers represent the duplication of all genes in pairs. It may be assumed that no thoroughbred genotype, not even that of St. Simon, has ever been absolutely flawless, as some important genes are always held in single strength in combination with undesirable recessive genes. Thus both layers of the raincoat have some tears here and there, but the holes do not correspond and so the rain does not penetrate. The animal appears to be perfectly sound. When a mating takes place one layer of the sire's raincoat is peeled off and united, in the offspring, with a layer of raincoat peeled off from the dam. Each layer has some tears but, provided that no close inbreeding is involved, the chances are that the tears will again not correspond and the new raincoat will still be waterproof. When an inbred mating takes place, on the other hand, the two double raincoats concerned are similar and are likely to have some of the tears in the same place. For this reason there is a danger that the tears will correspond in the two layers handed on to the offspring, and the wearer of the new raincoat will get wet. In other words, the offspring is liable to have some crippling unsoundness.

If the defective animals so produced are weeded out and not used for breeding, and inbreeding is continued with the sound progeny in each succeeding generation, the line will be purified gradually of the undesirable recessives and will breed true for the valuable characters. The closer the inbreeding, the quicker the process of purification will be. Every outcross resorted to will retard the process by the introduction of fresh impurities.

*Assessing the Risks of Inbreeding* (*I*)

The breeder contemplating an experiment in inbreeding has to weigh the risks against the advantages he may reasonably expect to gain. In this respect the pronouncements of some geneticists on the specialised subject of the thoroughbred have been singularly unhelpful. For example A. L. Hagedoorn wrote in his otherwise invaluable *Animal Breeding*: 'One of the most promising schemes in racehorses is the duplication of a famous speedy sire. This can be worked in the horse, because of the great age to which studs keep breeding. My idea would be to breed back a few daughters of such a male to their father, mate the mares so produced to the "grandfather", and continue as long as possible or until success is attained. This is a system of breeding for speed which has a sound genetical foundation, and the same thing cannot be said for the fanciful systems of "blending", in which some breeders seem to have faith.'

The passage implies so gross an oversimplification of the problems facing the breeder of racehorses as to divorce theory totally from practice. Many of the most vital questions have been begged. What is meant by 'speedy'? How is 'success' to be judged? If speedy is used in its normal connotation of a fast, non-staying animal, the end-product of an inbreeding process such as that recommended by Dr. Hagedoorn would quite probably be deficient in the stamina necessary to stay even the minimum distances.

That is just one of the pitfalls inseparable from the Hagedoorn system. The gravest danger of all follows from the fact that the racehorse has to be a physical and nervous machine perfect in all respects. This is not equally true of most farm animals and poultry which are valued for single qualities like milk yield or egg production. No cow is exposed to the same physical stresses as a runner in the Derby, and the excitement that the cow feels at the sight of a child picking buttercups in its field or the sound of a lorry passing down the lane is not to be compared with the nervous tension of a three-year-old colt during the lengthy preliminaries on Derby day. The weaknesses of a genotype are ruthlessly exposed by the racecourse test. Since some of the genes carried as recessives control such characters as infertility, roaring, blood-vessel breaking, unsoundness of limb and nervous trouble, it is obvious that the racehorse breeder who breeds in may be taking considerable risks.

Stallion fees, the price of shares in stallion syndicates and the overall costs of producing thoroughbred foals are so high that it is not surprising that many breeders are reluctant to accept the additional burden of producing a proportion of worthless offspring which inbreeding may involve. About 30 per cent of all thoroughbred matings in England and Ireland are infertile, and the risk of increasing this percentage by inbreeding often seems unacceptable. 'Better play safe by outbreeding and not take chances by inbreeding' is a fairly common reaction to the problem.

### *Assessing the Risks of Inbreeding* (*II*)

It must now be clear that success in inbreeding depends on the choice of parents who are free from serious physical and nervous faults. Stallions and mares who are themselves unsound are unsuitable material for inbreeding, but the problem goes deeper than that. Sound animals, as we have seen, may carry hidden faults as recessives, and for this reason it is advisable to examine their pedigrees for several generations back before deciding whether they are fit for use in inbreeding experiments. If any near ancestors have possessed serious hereditary flaws, then persistence in inbreeding is asking for trouble.

This point can hardly be overemphasised. An unsound sire is bound to pass his unsoundness on to some of his progeny if this character is due to a flaw in his genotype. Some even of his progeny who do not show the character will carry it as a recessive, which is liable to be disclosed by inbreeding.

From time to time there is a horse of exceptional merit who appears to be a worthy subject for inbreeding but for a single serious inheritable flaw. In a case of this kind the breeder has to weigh up the risk involved against the possible advantage of breeding in to him.

Blandford may be taken as an example. He was the outstanding Classic sire available in England or Ireland between the two world wars, and got the Derby winners Trigo, Blenheim, Windsor Lad and Bahram. He was an extraordinarily prepotent stallion. Unfortunately he was unsound, failed to stand prolonged training and broke down after winning the Princess of Wales's Stakes as a three-year-old. He transmitted his unsoundness to many of his progeny, though the fault was often hidden as a recessive. The crucial question for breeders was whether the attempt to perpetuate his excellent qualities, which included exceptional racing ability and perfect temperament, by breeding in to him was worth the risk that the process might uncover the unsoundness in the progeny.

In fact inbreeding to Blandford has been practised with success on numerous occasions. Espresso, who won the Grosser Preis at Baden Baden twice, was by Acropolis (by Donatello II, by Blenheim by Blandford) out of Babylon, by Bahram (by Blandford). Gallant Man, the winner of the Belmont Stakes in 1957, was by Migoli (by Bois Roussel out of Mah Iran, by Bahram) out of Majideh, by Mahmoud (by Blenheim). Mated with Rivaz, whose dam Mumtaz Begum was by Blenheim, Gallant Man got Spicy Living, one of the best three-year-old fillies in America in 1963. This was an instance of sustained inbreeding to Blandford, though it is only fair to add that the brilliant race- and brood mare Mumtaz Mahal also appeared three times in the pedigree of Spicy Living in the same number of generations as Blandford.

*Linebreeding*

Linebreeding is merely a less intensive form of inbreeding, and the frontier between the two is undefined. In *Animal Breeding* Hagedoorn related a conversation he once overheard at a poultry show: A young man had asked what the difference was between inbreeding and linebreeding. An older one answered: 'Well, son, it's this way: If you keep on breeding with your own birds, and you are successful, you speak of linebreeding. But if your results are bad, you can blame it to inbreeding.' Breeders of racehorses tend to differentiate between the two forms in much the same way.

Linebreeding is less effective than inbreeding in that the genotype of the chosen ancestor is not so well preserved. Extraneous genes are introduced at each generation and for this reason the approach to genetic purity, the possession of all important genes in double strength, is slower. On the other hand linebreeding is a safer proposition for the ordinary breeder who cannot afford the wastage which may result from inbreeding, because the chances of uncovering harmful recessives are reduced.

*Heterosis*

Heterosis, or hybrid vigour as it is otherwise called, is the phenomenon of increased vigour which sometimes occurs in the offspring when two genetically different animals are mated. One of the best examples of hybrid vigour is the mule, which is a cross between a horse and a donkey, usually between mares and large donkey stallions. The mule normally possesses the toughness, intelligence and longevity of the donkey, and the superior size, strength and speed of the horse. What happens is that the mule's genotype combines characters derived from each parent, and the combination makes it more efficient than either parent for certain kinds of work, especially pack transport.

Within the breed of the racehorse heterosis is most likely to be found when individuals representing two separate inbred lines are mated. In the process of inbreeding certain valuable characters are liable to be developed while others are neglected or left out altogether, but it is improbable that the same characters will be neglected in two separate inbred lines. For this reason the two lines may be complementary, the one holding in enhanced form characters which the other lacks, and vice versa, so that a mating of a stallion of one line with a mare of the other has the chance of producing superior offspring combining the virtues of both. This is what seems to have happened in the case of Hethersett, who was by Hugh Lupus out of Bride Elect. Hugh Lupus was inbred to one prepotent sire, Tourbillon, as he was by Djebel (by Tourbillon) out of Sakountala, by Goya II (by Tourbillon); and Bride Elect was inbred to another prepotent sire, Blandford, being by Big Game (by Bahram, by Blandford) out of Netherton Maid by Nearco, out of Phase by Windsor Lad (by Blandford). Blandford did not appear in the pedigree of Hugh Lupus, nor did Tourbillon appear in the pedigree of Bride Elect. Hugh Lupus won the Irish 2,000 Guineas and the Champion Stakes, but had his limitations as a racehorse. Bride Elect was fast but did not stay. Hethersett surpassed both his parents as, although he could not act on hard ground and was brought down in the Derby of 1962, he was a very high-class horse indeed when he won the St. Leger.

### *A Great Inbreeding Experiment*

The question of inbreeding in the thoroughbred has been obscured by the mass of prejudices and misconceptions which have surrounded the subject. Some writers have not facilitated an objective approach to the problem by using the word 'incestuous', with its overtones of moral condemnation, to describe matings which involve a close degree of inbreeding. Although better understanding should follow from a grasp of the way heredity works, it is not at all easy to relate the scientific principles to the realities of breeding the racehorse. For this reason it should be helpful to investigate one of the most concentrated experiments in inbreeding ever undertaken in connection with the thoroughbred and discover whether it throws any light on this specialised aspect of breeding. This experiment was carried out by the 12th Earl of Derby, the man after whom the Epsom Classic race was named, who used his 1787 Derby winner Sir Peter (also called Sir Peter Teazle), Sir Peter's sire Highflyer and grandsire Herod as the basis of his inbreeding.

Derby must have been induced to embark on this inbreeding experiment by his pride in Sir Peter, but was taking grave risks. It is true that Sir Peter himself had no discernible flaw. He was the outstanding racehorse of his day, was sound enough to win seventeen races before he broke down, and was a brilliant sire, getting four winners of the Derby, two of the Oaks and four of the St. Leger; but Herod, though an exceptional racehorse and sire, also had the serious drawback of being a blood-vessel breaker. When running as an eight-year-old against four opponents in a subscription purse at York in August 1766, Herod broke a blood-vessel in his head which, in the words of the contemporary account, 'caused him to be taken seriously ill, and prevented his coming in a better place than last'. The researches of J. B. Robertson led him to conclude that Herod was the principal source of blood-vessel breaking in the thoroughbred and that many of his descendants carried this weakness as a recessive.

Fortunately Highflyer was not a blood-vessel breaker. Foaled in 1774, he was the best racehorse of his day, and was never beaten and never paid forfeit. He was the founder of the racing fortunes of the Tattersall family and, as we have seen, his record as a sire compared favourably with that of Eclipse.

The first stage of Derby's experiment had Sir Peter and Sir Peter's three-quarters sister Wren as the ingredients. Sir Peter was by Highflyer (by Herod) out of Papillon, by Snap, and Wren was by Woodpecker (by Herod) out of Papillon, by Snap. Like Sir Peter, Wren was a good tough racehorse, and won no less than fifteen races. Sir Peter and Wren were mated four times between 1795 and 1801, and the four offspring all possessed some merit. They comprised Robin Redbreast, a winner of four races including the King's Plate at Warwick; Agonistes, a winner of fifteen races including a sweepstakes of 1,000 guineas; Tiney, a filly whose name presumably was descriptive, who ran only once but proved a valuable brood mare, producing twelve foals; and Milo, who won the Earl of Chester's Plate at Chester.

In 1794 Derby played a variation on the same theme by mating Wren with a son of Herod, Phenomenon. The product was Bellissima, who showed even better form than any of the products of the Sir Peter – Wren mating, and won the Oaks. Bellissima afterwards bred five foals, none of them in her own class. Phenomenon had the curious distinction of being last in the Derby and first in the St. Leger, and sired one other Classic winner besides Bellissima, the St. Leger winner Ambidexter.

The second stage of Derby's inbreeding took place from 1799 to 1805. During this period Sir Peter was mated with his full-sister Brown Bess five times. Of the five foals that resulted, the fillies Maud and Margaret showed the highest racing class; Maud was third to Pelisse in the Oaks in 1804 and Margaret was second to Briseis in the Oaks three years later. Soon after she had run in the Oaks, Maud was sold and taken to Ireland, where she ran in heats over the aggregate distance of twenty miles in eight days, and came out the next season to win two more races. Margaret won a race at Lichfield after running in the Oaks. Another product of the Sir Peter – Brown Bess mating, the colt Jacobus, won a race at Chester.

Maud, who had been re-named Pandora, broke a leg in a race at Tuam and presumably was destroyed. No returns for Margaret were made to the General Stud Book, but there is evidence that she had at least two foals. One, the colt Momentillo, won three of the four races in which he took part; the other, the filly Ulvira, ran four times but did not win. Brown Russet, a full sister of Maud and Margaret, had normal fertility and produced seven foals between 1807 and 1816.

The third stage of the experiment had Milo, one of the successful products of the Sir Peter – Wren cross, and a Whiskey mare as its principals. This mare was by Whiskey (by Saltram out of Calash by Herod) out of Amelia, by Highflyer. The effect of this mating was that any product would have Herod in four of the eight male places in the fourth generation. The mating was repeated ten times, and was fertile on five of those occasions. Unfortunately the results were not nearly as satisfactory as in either of the previous stages. Two foals were still-born, one colt died as a two-year-old, another colt was exported without trace, and only Rinaldo, a colt foaled in 1810, was submitted to the racecourse test in England. Rinaldo was temperamental as a three-year-old, when he bolted off the course at Chester and swerved so badly in another race that he went the wrong side of a post. The following year he did much better and won the Gold Cup at Preston, then an important race. Rinaldo won 13 races worth £1,429 altogether.

Milo sired a total of 79 foals between 1808 and 1826; of these foals 47 ran and won 103 races, including walk-overs and matches, of £10,858. This was the record of a moderate stallion, and it must be remembered that he was not a top class racehorse. However he did make one contribution of profound significance to thoroughbred evolution. His second best son after Rinaldo was Eryx, the winner of seven races worth £766. Eryx, whose dam was inbred to Herod in the third generation, was used as a stallion by Lord Derby. He had only seventeen recorded foals, but one of these was Elvira, who was exported to France and became the dam of Lanterne, the winner of the French Derby and the French Oaks in 1844. A quarter of a century later the family returned to England when Fenella, a great granddaughter of Lanterne, was imported by William Blenkiron, the owner of the Middle Park Stud and one of the first breeders to operate commercially on a large scale. This Fenella was the direct female line ancestress of Sansovino and Hyperion, whose victories in the Derby for the 17th Earl of Derby were a brilliant climax to family fortunes both equine and human without parallel in the annals of the Turf.

The intensive inbreeding practised by the 12th Earl of Derby triumphed in the long run—in the very long run. But his methods in the later phases, using horses like Milo and Eryx, could never be advocated as rational breeding policy. This whimsical perseverance with second-rate material reduced his own stud to a low ebb, and the chain of heredity linking Milo to Hyperion is too long and too tenuous to bring him any credit.

Derby certainly took a chance by inbreeding to Herod on account of that horse's recessive gene affecting the breaking of blood vessels, but he seems to have got away with it in this respect. Otherwise his choice of material in the earlier stages was faultless, especially in his reliance on such admirable horses as Sir Peter, Highflyer, Phenomenon and Wren. From the point of view of racing merit the results of the Sir Peter – Wren, Phenomenon – Wren and Sir Peter – Brown Bess matings were impressive. The ten matings known to have taken place produced one Classic winner, two other animals of Classic standard and four other animals with good winning form. Moreover at least three of these winners possessed exceptional ability to stand up to hard work. This proportion of good form allied to toughness would not be matched by a sample of ten matings chosen from the General Stud Book at random, nor even by a random sample of ten matings involving a stallion as excellent as Sir Peter.

These examples show that the harmful effects attributed by prejudice to inbreeding are not inevitable. Inbreeding does not necessarily cause loss of fertility or vitality, debilitating nervous trouble or general degeneracy. On the other hand, Derby's experiments did not improve his bloodstock and were not accompanied by the rigorous culling of inferior progeny which would have given them real scientific value. In the end the 12th Earl of Derby contributed a romantic chapter to thoroughbred history; he did not extend the frontiers of knowledge in the way that a more serious and dedicated breeder might have done with similar opportunities.

*Back to St. Simon*

We must now retrace our steps to St. Simon and investigate whether inbreeding to him, or in other words partial preservation of the completed jigsaw puzzle represented by his genotype, has brought significant benefits to the breed of the racehorse.

When a sire stands out above his contemporaries to the extent of St. Simon, a certain amount of inbreeding to him is inevitable. So many of the best stallions and mares of the next few generations will bear his name in their pedigrees that breeders will scarcely be able to avoid mating some of them in their efforts to continue producing animals of the highest class. This has happened so often in the case of St. Simon that it would be difficult to find a good racehorse of the present time whose pedigree does not contain the name of St. Simon several times. Indeed St. Simon appears no less than thirteen times in the pedigree of Ribot, the Italian-bred horse whose unbeaten record of sixteen victories, including the Prix de l'Arc de Triomphe twice and the King George VI and Queen Elizabeth Stakes, marked him as one of the great racehorses of all time. There is an individual inbred to St. Simon in each of the quarters of Ribot's pedigree, these being Havresac II (2 × 3), Apelle (4 × 3), Pharos (4 × 3) and Papyrus (4 × 3). Although St. Simon does not appear until the sixth generation of the pedigree of Ribot, he represents a high proportion of the whole if individual ancestors are regarded as units, and a still higher proportion if the prepotency of St. Simon is taken into account. In order to find examples of successful inbreeding to St. Simon we may turn to one of the greatest brood mares and one of the greatest racehorses and sires of the twentieth century, Friar's Daughter and Nearco respectively.

*Friar's Daughter*

Friar's Daughter, foaled in 1921, was a brown by Friar Marcus (by Cicero out of Prim Nun by Persimmon, by St. Simon) out of Garron Lass by Roseland (by William The Third by St. Simon), out of Concertina by St. Simon. Thus her dam Garron Lass was inbred to St. Simon (3 × 2), and her sire Friar Marcus introduced a third cross of St. Simon within the first four generations of her pedigree.

Friar's Daughter was bought as a yearling on behalf of the Aga Khan for 250 guineas. She was one of the best thoroughbred bargains ever made. In 1922 the Aga Khan was beginning to buy the well-bred fillies that were destined to be the foundation of his stud and stable. He bought Mumtaz Mahal for 9,100 guineas the same year. Mumtaz Mahal was more than 2 stone better than Friar's daughter when they were tried together at home. There was not so much to choose between them when they went to stud; the immediate offspring of Friar's Daughter were the better, but the influence of Mumtaz Mahal on succeeding generations was more profound.

Friar's Daughter raced only as a two-year-old, when she won one race worth £168 over five furlongs at Alexandra Park. She was second in four other races and was third once. Thus the preservation of the St. Simon genotype did not produce a good racehorse in her case, though the cause of her moderate form was probably not any shortcomings in this respect, but rather was environmental. She had been seriously ill as a foal and was saved only by the constant care of her then owner Mrs. Plummer. The wonder was not that her form was moderate but that she was capable of winning a race at all.

While adverse environmental factors may prevent a horse realising the excellence of his genotype in terms of his own racing form, they do not cause any deterioration of the genes that are transmitted to his offspring. Thanks to the inbreeding to St. Simon, Friar's Daughter had many of the genes that make for excellence in the racehorse in double strength, and was a prepotent brood mare for this reason. She produced a foal in thirteen of the first fourteen seasons in which she was covered; then, at the age of nineteen, she became infertile and never conceived again. Of her thirteen foals eleven were winners in one part of the world or another. The most famous of her offspring was Bahram (by Blandford), who was unbeaten in nine races and gained the 'Triple Crown' of victories in the 2,000 Guineas, Derby and St. Leger. Her son Dastur (by Solario) was second in each of the Triple Crown races, besides winning the King Edward VII Stakes, Irish Derby, Sussex Stakes and Coronation Cup and running a dead-heat with Chatelaine for the Champion Stakes. Several of the offspring of Friar's Daughter were successful abroad and Sadri, a full-brother of Dastur, won the Durban July Handicap, one of the most important races in South Africa.

*Nearco*

Nearco was by a sire, Pharos, who was inbred to St. Simon 4 × 3, and had as his maternal grandsire a horse, Havresac II, who was inbred to St. Simon 2 × 3. Thus St. Simon appears twice in the fourth generation and twice more in the fifth generation of the pedigree of Nearco.

In some respects Nearco, foaled in 1935, resembled St. Simon more than any of St. Simon's own sons or any other horse inbred to St. Simon. He possessed much of the same 'electricity', and as a racehorse asserted the same absolute supremacy over all other horses of his own time. He was never extended in any of his races. The note about him written by his breeder Federico Tesio in his private stud book sums him up concisely: 'Beautifully balanced, of perfect size and great quality. Won all his fourteen races as soon as he was asked. Not a true stayer, though he won up to 3,000 metres (Gran Premio di Milano and the Grand Prix de Paris). He won these longer races by his superb class and brilliant speed.'

His victories were equally divided between his two-year-old and three-year-old seasons, and were over distances from five furlongs upwards. If his racing had been confined to his native Italy he might have gained his reputation cheaply, but in the Grand Prix he decisively defeated the English Derby winner Bois Roussel and the French Derby winner and second, Cillas and Canot.

Nearco spent his whole stud career, which lasted from 1939 to 1957, in England. He was leading sire of winners twice, in 1947 and 1948. In this respect he compares unfavourably with Hyperion, the other great sire of the mid-twentieth century in England, who was leading sire six times. On the other hand Nearco and Hyperion were in the top ten sires the identical number of times, which was sixteen. That total has been exceeded only once since the middle of the nineteenth century, and that was by St. Simon himself, who was in the top ten sires seventeen times. One of the most striking features of the stud career of Nearco was his marvellous consistency, for he was in the leading eight sires for fifteen consecutive years, from 1942 to 1956.

This consistency testifies eloquently to the prepotent quality of Nearco. The high class of many of his progeny was just as remarkable. He sired the 2,000 Guineas and Derby winner Nimbus, the Derby winner Dante, the Oaks winners Neasham Belle and Masaka, the St. Leger winner Sayajirao and a host of horses who achieved fame in one category or another, including Nasrullah, Mossborough and Royal Charger. Inbreeding to him has been extremely effective. His name is duplicated in the pedigree of the 1972 Derby winner Roberto and triplicated in the pedigree of the 1969 Derby winner Blakeney; and in the single season of 1973 inbreeding to Nearco was a feature of the pedigrees of Rheingold, Kalamoun, Hippodamia, Sun Prince, Habat and Giacometti, who all succeeded at a high level of European competition.

*Conclusion*

Inbreeding has a place in modern racehorse breeding as a means of capitalising on the production of a great performer and raising thoroughbred ability on to a new plateau of achievement. The analogy with a plateau is precise, because progress within a system of inbreeding is limited to the most favourable arrangement of the genes possessed by the individual or individuals to whom the inbreeding is taking place. Once that arrangement has been obtained, further progress can be achieved only by outcrossing, introducing fresh genes and securing the renewed vigour that may spring from pedigree contrast.

Inbreeding for inbreeding's sake will serve no purpose, and the breeder who decides to breed in must keep his aim clearly in mind. This aim should be to preserve or reconstitute the genotype of an outstanding individual, whether horse or mare, who proved his ability to pass on his own desirable characters to his progeny; and a condition of success is that the individual concerned should have had no serious heritable defects, or at any rate should have transmitted those defects only to a limited number of his progeny. Questions of delicate calculation, of weighing of pros and cons, are involved.

# 7

# *Some Aspects of Selection*

WE HAVE SEEN THAT SELECTION of suitable stock is essential for successful inbreeding; indeed judicious selection, based on the characteristics of the animals available and the aims of the breeder, is of the first importance in every mating, whether inbreeding or outbreeding is involved. In some respects the task of the breeder who has decided that inbreeding is the answer to his immediate problem is simpler, since his decision has automatically restricted his field of choice. What the criteria of selection in the breed as a whole should be is a question that has given rise to a great deal of controversy, and no study of thoroughbred breeding would be complete without an examination of it.

The famous trainer Atty Persse, who was responsible for developing the brilliance of The Tetrarch and lived to the age of ninety, is said to have been asked by a fellow guest at a dinner party while on a visit to America as a young man: 'Tell me, Mr. Persse, what do you consider the most important quality of the racehorse?' 'Speed,' was the reply. 'Yes indeed,' went on the interrogator, 'but tell me, what do you consider the next most important quality?' 'More speed', said Persse. 'Quite, quite,' pursued the American, becoming a little impatient, 'but after that?' 'Still more speed,' declared Persse without a moment's hesitation.

The story may be apocryphal. If it is true, Persse's answers pass muster as dinner-table judgements, born of no profound thought but intended to emphasise a point of view. As a set of principles upon which selection should bc based they will not do at all, because all the speed in the world may be nullified by unsoundness of heart or wind or limb; by lack of courage or bad nerves; by poor constitution; by inability to maintain speed for for the required racing distance; and by a variety of other causes. The Tetrarch himself did not run after his two-year-old season because he was not sound enough to stand training.

In his book *Horse Breeding Recollections*, published in 1883, Count Lehndorff, the manager of the German Imperial Stud, stated: 'The principal requisite in a good racehorse is soundness, again soundness, and nothing but soundness.' In his memoirs *Men and Horses I Have Known* George Lambton described Lehndorff as the 'finest judge of bloodstock in the world' and added that he had tried to follow Lehndorff's principles all his life. Nevertheless Lehndorff's dictum cannot be accepted as the whole truth any more than that of Persse. The horse on which the Duke of Wellington sits eternally at Hyde Park Corner is almost literally as sound as a bell of brass—the statue is of bronze—but, being completely immobile, is useless for racing purposes.

It must be clear that selection should be based not on a single quality to the exclusion of all others, but on a combination of qualities which include speed, soundness, conformation, pedigree and constitution. The breeder's problem is to get his priorities right and ensure that no quality whose omission may defeat his purpose has been left out. The problem resembles one of those prize competitions in which the entrants have to place various properties of a new car or piece of domestic hardware in their correct order as judged by a panel of experts. In thoroughbred breeding the racecourse test, not a panel of experts, decides the successful candidates.

There is little doubt that Persse was right in asserting the priority of speed, although he erred in overemphasising this single aspect of a complex subject. The priority of speed may be deduced from any acceptable definition of a racehorse. Races are not won by staying in the same place. The value of speed was appreciated by the Duke of Wellington himself, for the horse represented in the Hyde Park Corner statue is Copenhagen, a winner of a match at Newmarket and a sweepstakes at Huntingdon, whom he rode at Waterloo. Fortunately the result of the battle saved the Duke from the need to take advantage of his charger's speed.

The Aga Khan, who bred the Derby winners Mahmoud, Bahram and Tulyar and built up one of the greatest studs of all time between the two world wars, was convinced of the primacy of speed. In an article in *The Times* published in June 1950, when the frequent successes of French-bred horses in the English Classic races were spreading despondency among breeders and the whole racing community in England, the Aga Khan wrote: 'Breeding for speed has made the British Thoroughbred, and the ultimate remedy must be more and more speed.' He believed that any attempt to downgrade speed, or to elevate any other qualities to equal status, would reverse the process by which the British Thoroughbred has become the ancestor of all thoroughbreds, including the French. He concluded the article: 'I have only one piece of advice to offer—be careful when you throw out the water from the tub; do not let the baby fall as well. And that baby is speed.'

The Aga Khan thus insisted on the priority of speed, but certainly did not advocate that it should be prized to the exclusion of other desirable and essential qualities. The races which bear the greatest prestige and the greatest monetary rewards in nearly all countries in which the Turf flourishes are run over distances from a mile to a mile and three-quarters, so that a breeding policy which involved the use of stallions and mares capable of staying only sprint distances in the best company could have only limited objectives. The complicated relationship of speed and stamina needs to be understood. Any fit thoroughbred can gallop a mile, or one and a half miles, or two and a half miles, if he is allowed to do so in his own time, and the question of his stamina is concerned with the distance over which he can race most effectively against animals of similar ability—in other words, the distance of his best performance. A champion sprinter, unable to stay more than six furlongs in his own class, could beat a staying selling plater over two miles, because the selling plater would never get him at full stretch. Even in the highest class of racing the rare horse is found who is able to beat the best of his contemporaries over all distances, including middle and long distances, not because he stays as well as they do but because he is so fast that he does not have to exert himself to beat them.

We have seen, for example, that Nearco was able to beat the English and French Derby winners in the Grand Prix although he was not a true stayer. If Nearco had had to compete with another horse of his supreme class his distance of maximum effectiveness would probably have been revealed as one and a half or one and a quarter miles. But horses of Nearco's class can be numbered on the fingers of two hands for the whole of thoroughbred history, and his staying power was never put to the test. Not surprisingly the Aga Khan made the fullest use of the services of Nearco when he went to stud, and obtained such notable horses as Masaka, Nasrullah, Rivaz and Hafiz II as a result. Sir Ivor and Nijinsky were later examples of horses who were able to stay 1½ miles—the 1 mile 6 furlongs and 127 yards of the St. Leger in the case of Nijinsky—without being true stayers, and it is significant that they were both male line descendants of Nearco.

It is tempting, and labour-saving, when classifying racehorses, to include all winners of the Derby and other principal prestige-bearing middle-distance races in the same category as Nearco, but in fact the general description of 'Classic winner' comprises significant variations which ought not to be ignored. The vital line of demarcation should be drawn, perhaps, to separate those middle-distance Classic winners who are capable of beating the best sprinters of their day over five and six furlongs, from those who are not. Those Classic winners who are faster than the best sprinters represent the ideal combination of speed and stamina and may be expected, above all others, to make a valuable contribution to the development of the breed when they go to stud.

Nearco is a perfect example of this type of horse. St. Simon is another; admittedly he did not win a Classic race because he was unable to run, but he beat the best stayers in the Ascot Gold Cup and had incomparable speed. Hyperion had the speed to win the New Stakes at Royal Ascot as a two-year-old and was one of the fastest horses of his age, and won the Derby and St. Leger the next year. Bahram, two years after Hyperion, won the National Breeders Produce Stakes, the Gimcrack and the Middle Park Stakes and was placed top of the two-year-old Free Handicap well above Bellacose; the next year Bellacose became champion sprinter, while Bahram was otherwise engaged winning the Triple Crown. Nearco and Hyperion became the greatest Classic sires of the middle years of the twentieth century. Bahram got the St. Leger winner Turkhan, the 2,000 Guineas winner Big Game and the Coronation Cup winner Persian Gulf during his brief stay at stud in England, though his later career suffered from his inability to adapt himself to changes in climate and environment when he was transferred first to the United States and then to Argentina.

The argument here is not that it is necessary or even desirable for Classic horses to race over short distances as two-year-olds. That is another question altogether. The point at issue is simply that horses like Hyperion and Bahram actually demonstrated that they were as fast or faster than the best sprinters of their age. The Aga Khan's metaphor of the baby and the bath tub must be interpreted in the light of these factors. Indiscriminate attempts to breed for stamina and middle-distance ability, without due regard to the preservation of speed, are liable to defeat their own object and lead to the production of slow horses and mediocrities.

Where the aim is merely to breed speedy animals the problem is, of course, greatly simplified; in this case the breeder needs only to take care that he does not pursue this aim with such singleness of purpose that ability to stay even the minimum racing distances is lost.

### *Racing Ability and Breeding Value*

Horses like St. Simon, Hyperion, Bahram and Nearco are the most highly valued for stud purposes at the time they retire from racing. Breeders assume that they will transmit their own exceptional qualities to their progeny, but this is a matter of probability and not of certainty. Many good racehorses, though few great ones, have been disappointing sires, and a number of other horses, among whom Phalaris and Hampton may be chosen as shining examples, have achieved great distinction at stud after failing to reach the top class on the racecourse. The only way of making sure of a horse's value as a stallion is by waiting to see how he breeds—that is, by the test of the ability shown by his progeny. Unfortunately the progeny test, though also the acid test, can never be fully applied in the case of the thoroughbred. To be judged fairly, a stallion should have had three crops of foals to race until they are three and four years old, and by then, assuming that he stayed in training until he was four, he has reached the age of twelve and is nearly halfway through his expectation of useful stud life.

The problem is different in the case of, say, poultry or dairy cattle, in which the cock or the bull does not himself perform the functions for which his progeny are valued. The progeny test assumes crucial importance when it is a matter of finding out whether a cock gets hens who lay well or a bull gets cows with a high milk yield. The large majority of thoroughbred stallions have done the same thing that their progeny are called on to do, which is to race; and if a horse has been a high-class racer he must possess most of the right genes and should pass them on to some, at least, of his progeny. For this reason the breeder of racehorses is able to take short cuts which the breeders of many farm animals would have difficulty in justifying.

The racecourse test is thus the first and indispensable guide to breeding value, but there may be cogent reasons for upward or downward revision of the original valuation during an animal's stud career in the light of the evidence provided by his progeny. The prime requirement is to avoid hasty judgements and the illogical swings of fashion to which the bloodstock market is prone.

There are various statistical aids to the assessment of the relative success of sires, and so to their breeding value. These consist mainly of the annual lists of winning sires, which show the number and value of races won by the progeny of each; fertility percentages (i.e. the percentage of mares covered by each stallion that become pregnant); and the average winning distance of the progeny. The average winning distance, though based on the victories of three-year-olds and upwards, tends to underestimate the potential stamina of most of the progeny because it takes into account victories gained over distances well short of the limit of stamina. For example, Alycidon, who was able to stay two miles five furlongs, won over a mile as a three-year-old.

The limitation of these methods of assessment is that neither individually nor collectively do they relate achievement to opportunity. In an attempt to obtain a truer picture than that given by money values of races won as the principal criterion, Mr. J. A. Estes, on behalf of the Research Bureau of the American Thoroughbred Breeders Association, devised what is known as 'the average earnings index system'. His formula for calculating the index of any sire in a particular season is:

$$\frac{\text{The total earnings of the sire's progeny for the season}}{\text{Average earnings for every horse in training} \times \text{total runners by him}}$$

In other words, the index is the product of actual earnings divided by expected earnings. It is claimed that a cumulative index covering several seasons will provide a sound base for comparing one sire with another irrespective of changes in the value of money.

One may agree that the average earnings index system should help to do justice to worthy sires represented by small numbers of runners, just as the true merit of these sires may be concealed by a system which takes account only of total earnings. On the other hand the average earnings index system may solve one problem only to pose and leave unanswered other and equally important questions. Why has the sire who showed up better under the average earnings index system than by the criterion of total earnings had so few runners? If the cause is poor fertility, then this detracts seriously from his breeding value. If his fertility is normal, is the cause high incidence of unsoundness, or weak constitution, or lack of racing ability among his progeny? Any of these causes would again amount to a serious reflection on his breeding value. If, on the other hand, the cause is a high rate of exports among his progeny or inadequate patronage by owners of brood mares, then the average earnings index system may indeed perform a useful service by drawing attention to his merits.

Hugh Lupus and Solar Slipper are two sires who have had the best index figure for a season, while finishing well down the list of successful sires by money earned. In 1962 Hugh Lupus had an index figure of 7·27, which gave him a long lead over the second sire by this criterion, Never Say Die, with an index figure of 4·76. In the ordinary list of winning sires Never Say Die was first, his progeny having won £65,902, whereas Hugh Lupus was only seventh with a total of £43.151. Never Say Die was responsible for eighteen individual winners from forty-two runners, including the Derby winner Larkspur, but Hugh Lupus had only five other winners from eighteen runners besides the St. Leger winner Hethersett. The reason for the discrepancy in the position of Hugh Lupus in the two rankings was quite simply low fertility during his early seasons at stud; his fertility percentages for the three covering seasons from 1957 to 1959 were 27·27, 60·87 and 42·42, so his index figure was inflated to a flattering extent by the earnings of Hethersett. It is only fair to add that his fertility for the next covering season rose to a more acceptable 67·57 per cent.

In 1961 Solar Slipper had the best index figure of 6·66, and Aureole had the second best index figure of 6·27. By contrast Aureole was top of the ordinary winning sires list, his progeny having won £90,898, whereas Solar Slipper was only tenth with stakes earnings of £28,153 credited to his progeny. Aureole was represented by fifteen winners from forty-eight runners including the St. Leger winner Aurelius, the Eclipse Stakes winner St. Paddy and the Timeform Gold Cup winner Miralgo, but Solar Slipper had only four other winners from fourteen runners besides the Ascot Gold Cup winner Pandofell. The reason for the discrepancy in the positions of Solar Slipper in the two rankings was that he had been exported to America in 1956 and had only four-year-olds and older horses to represent him by 1961. As a result the success of Pandofell, the best horse he had got apart from the 1955 Irish Derby winner Panaslipper, inflated his index figure for the 1961 season artificially.

It can hardly be claimed that the average earnings index system would have been of much practical value to breeders in the cases of either Hugh Lupus or Solar Slipper. Special factors inflated the index figure in each case. A more general conclusion may be that the average earnings index system implies too narrow an interpretation of 'opportunity' to be a satisfactory method of judging the breeding value of a stallion. The only kind of opportunity worth bothering about in this connection is a matter of the quantity and quality of the mares he has covered. The production of a valid success-opportunity index by which different stallions might be compared would involve detailed analysis of the merits of thousands of mares and would not be a feasible operation. Nevertheless any fair verdict on the achievements of a stallion must be based on an estimate of the overall standard of his mates, even though that standard cannot be expressed in the form of an index figure.

### *The Correlation of Racing Merit and Breeding Value in Mares*

In this study of selection the focus so far as been on stallions, but the arguments presented apply with equal force to the role of the brood mare. What is sauce for the gander is sauce for the goose. Indeed it was the researches of J. B. Robertson among the records of mares that gave an empirical endorsement to scientific theory in this particular field; for Robertson was able to establish, as he wrote, 'a high correlation between racing merit in mares and the power to produce offspring which can not only race, but race with distinction'. Robertson discovered that, of 1,000 mares chosen from the General Stud Book at random, 66 per cent never ran or showed no form. He then turned his attention to the Classic winners of the fifty-year period from 1862 to 1913, and discovered that only 28 per cent of their dams never ran or showed no form. On the other hand, 50 per cent of the Classic winners were out of mares who had considerable racing ability, and 26 per cent were out of mares who had great racing ability. The last figure is impressive, because Robertson classified only 3 per cent of the mares in the random sample as mares of great racing ability. Even when due allowance has been made for the fact that good racemares tend to be mated with better stallions than a random selection of mares, the results of Robertson's researches seem conclusive, and permit the deduction that, whatever exceptions to the rule may have cropped up from time to time, animals of high racing merit have better expectations of success at stud than those who have shown no form or only moderate form.

### *Limits to the Application of this Rule*

Although there is a correlation between racing and breeding merit in mares, it would be impossible to restrict entry to stud to those mares who had actually shown good form on the racecourse. In accordance with Robertson's principle of random selection, the records of twenty mares whose first produce were registered in Volume XXIX of the General Stud Book have been traced through to the end of their stud careers for the purposes of the present study. This exercise revealed that those twenty mares had a total of 139 living foals, of whom 68 were fillies; in other words, they had an average of approximately 3½ living filly foals each. Thus it would have been necessary for one in every 3½ of those fillies to go to stud simply in order to maintain the level of the brood-mare population. In fact this batch of mares overfulfilled its contract in this respect, as 37 out of the 68, or more than half the filly foals, were registered themselves as having had foals in due course. On the other hand, only 5 of the 37 were winners, so it is clear that no very strict tests of suitability can have been applied in selecting them for stud purposes.

A larger sample of mares might produce somewhat different figures, but it is most unlikely that the main inference, that the scope for selection of mares is very limited over the breed as a whole, would be invalidated. By contrast, it is possible to apply strict standards to the selection of male thoroughbreds for stud. The large majority of stallions have shown good racing form; those who have not are mostly very well bred or have failed to reproduce good home form in public owing to unsoundness or accident.

Nevertheless it may be unwise to abandon a minimum standard of performance in the selection of mares for stud. Some successful breeders insist that a mare to be worth breeding from should either have won a race of some sort or have been prevented from doing so only by adverse circumstances. A small mare who has shown fair racing ability may be a good prospect for stud; unless she has been bred from a small parent or parents the lack of size which has limited her usefulness as a racehorse may have been due to some environmental influence, before or after birth, while the fact that she has been able to race reasonably well indicates that she has a satisfactory genotype. Catnip and Aloe were two famous brood mares who were small, yet not entirely without ability on the racecourse. Catnip was described as a 'light, narrow filly, that carried hardly any flesh', but she did manage to win a nursery worth £100 at Newcastle. She was sent by the executors of Major Eustace Loder to the Newmarket December Sales in 1915 and bought for 75 guineas by Federico Tesio, who took her back to Italy and bred from her Nogara, a winner of the Italian 1,000 and 2,000 Guineas and the dam of Nearco.

Unlike Catnip, Aloe did not have a single victory to her credit, but was really much the better racehorse of the pair. She was second to Pennycomequick, who afterwards won the Oaks, in the Haverhill Stakes, and was also second to Nuwara Eliya in the Nassau Stakes. Given an extra inch or two in height, and length and scope in proportion, Aloe would probably have been a high-class racehorse. Aloe produced two brood mares of note, Feola and Sweet Aloe. Feola herself was second in the 1,000 Guineas and third in the Oaks, and bred the 1,000 Guineas winner Hypericum; Angelola, the dam of Aureole; Knight's Daughter, the dam of Round Table; and Above Board, the dam of Doutelle and Above Suspicion. Sweet Aloe became the granddam of the St. Leger winner Alcide and great granddam of the Derby winner Parthia.

Although it is necessary to breed from many mares who have failed to win or have shown only moderate winning form, the examples of Catnip and Aloe suggest that the process of selection and rejection need not be entirely haphazard. Moreover Catnip and Aloe both ran ten times and thus proved themselves at least reasonably sound.

*The Relative Importance of Sire and Dam*

The controversy about the relative importance of sire and dam is perennial. Since a foal receives half of its make-up of genes from its sire and the other half from its dam, it is obvious that the two parents are potentially of equal importance, and a prepotent mare may transmit important characters to her offspring just as surely as a prepotent stallion. On the other hand, a good sire inevitably has a much wider influence than a good mare because he has so many more sons and daughters. A good stallion normally covers forty to forty-five mares every season; a good mare is seldom at stud for as many as half that number of seasons in her whole career, and can have only one foal each season unless she does her breeder the disservice of producing twins. St. Simon covered 775 mares during his time at stud; Sun Chariot, one of the greatest racemares of all time, was covered in only eighteen stud seasons during her whole career.

The other side of the question is that a good mare has a better chance of producing a high proportion of worthwhile offspring than a good sire, because she is always covered by a horse of a class equal to her own whereas a good sire is bound to cover many inferior mares. Sun Chariot's mates were Blue Peter, Big Game, Ocean Swell, Dante, My Babu, Tenerani, Tulyar, Pinza and Krakatao—all Classic winners with the exception of Krakatao, who was a high-class horse and won the Sussex and the Rous Memorial Stakes and the Chesterfield Cup. No stallion has ever received mates of such consistently high quality. Crepello won the 2,000 Guineas and the Derby and was considered a Classic horse of much above average merit. His reputation assured him of as high an all-round standard of mares as any stallion could recieve; yet an analysis of his eighty-two mates for the stud seasons of 1961 and 1962 reveals that they included only three Classic winners, these being Barbara Sirani (Italian St. Leger), Amante (Irish Oaks) and Festoon (1,000 Guineas). The remainder included twenty-one who may be classified as high-class winners, these being Be Careful, Collyria, Nagaika, No Saint, Panga, Plump, Queensberry, Toscanella, Abelia, Above Board, Camille, Crotchet, Duplicity, Kandy Sauce, Mirnaya, Rose of Medina, Sea Parrot, Star of India, Wake Up' and Urshalim (twice). The rest consisted of nineteen classified as good winners, and thirty-nine classified as small winners, non-winners and mares who never ran. In other words, of Crepello's mates during those two seasons only 3·7 per cent were Classic winners and 25·6 per cent high-class winners, and those two categories combined accounted for no more than 29·3 per cent of his mates. By contrast 100 per cent of Sun Chariot's mates were in these two categories.

These factors have to be kept in mind when assessing the stud achievements of a mare like Sun Chariot. She was barren three times, slipped her foal three times, and had one colt who died young. This left her with eleven surviving foals, of whom ten ran and seven won a total of eighteen races. Four of her offspring were high-class performers. They were the unbeaten Blue Train, whose three victories included the Newmarket Stakes; the Imperial Stakes winner Gigantic; the Rous Memorial and Sussex Stakes winner Landau; and Pindari, who won the Solario, Craven, King Edward VII and Great Voltigeur Stakes and was third in the St. Leger. Disappointment has been expressed that Sun Chariot did not produce an animal as good as herself, but no stallion has got such a high proportion of high-class performers as she did; but then again, no stallion has had such a high proportion of high-class mates. Generally speaking, the two kinds of opportunity, one based on quality and the other on quantity, cancel each other out, and tend to confirm that in individual matings the sire and the dam are potentially of equal importance.

### *Fine Blood and Rough Blood*

Since excellent performance on the racecourse is evidence of a good genotype, it follows that the best breeding policy must be to mate the best racehorses of both sexes provided that the conditions of soundness, constitution, temperament and compatibility of pedigree are satisfied. The principle stands, although it has to be modified in practice in accordance with two considerations, which are the necessity of breeding from many inferior mares in order to keep up numbers and the fact that a generally good genotype may be frustrated in action by environmental causes or by the lack of perhaps only one essential character. Any suggestion which violates this principle, like the so-called 'Chismon adage', should be regarded with suspicion.

William Chismon began his working life as a clerk in the Post Office, but acquired such a remarkable knowledge of the Stud Book that he became secretary to Colonel Hall Walker, afterwards Lord Wavertree, who bred the Classic winners Minoru,

Cherry Lass, Night Hawk and Prince Palatine and presented the National Stud to the Government in 1915. Chismon also acted as adviser to several other breeders. He wrote in 1916: 'The reasons why breeders select their stallions are often somewhat amazing. We have the owner of high class mares who books subscriptions only to the highest priced stallions. . . . His mares are probably very finely bred, and, because as a rule, the high priced stallions are just as finely bred, the breeder of this type is asking for trouble. He forgets that fine gold will not wear well without an alloy; neither will the progeny of his finely bred mares be very robust or sturdy if got by stallions whose pedigrees contain too much fine blood.'

Chismon's metaphor is inexact, since the degree of refinement or clarity of the blood is irrelevant. Although he did not use it, implicit in his remarks is the phrase 'rough blood', which is in general usage and is contrasted with 'fine blood'. 'Rough blood' and 'fine blood' are graphic phrases, but their indiscriminate employment in discussions of breeding problems is bound to lead to confusion because genes, not blood, are the bearers of inherited characters. The reason why some Clasically bred horses show weakness of temperament and constitution has nothing to do with an excess of fine blood, but is that their breeders have pursued the goal of top-class racing ability without having due regard to other essential factors. Indeed the single-minded pursuit of class is almost bound to lead to flawed products from time to time.

It sometimes happens that too high a concentration of dominant influences in a pedigree leads to physical or temperamental deterioration. John Hislop made the point succinctly in his story of the breeding and racing career of Brigadier Gerard,* winner of the 1971 2,000 Guineas and of 16 of his 17 other races. The parents of Brigadier Gerard were Queen's Hussar and La Paiva, and Hislop wrote of their union:

'Regarding the mating of La Paiva with Queen's Hussar on blood lines, the issue was not very complicated, since her pedigree contains remarkably few predominant influences in the early removes.

'The advantage of such a pedigree is that it offers great scope in the matter of sires, since there is little danger of arriving at an unbalanced mating, through concentrating too many similar influences in the pedigree, in particular, over-refinement.

* *The Brigadier,* by John Hislop. Martin Secker and Warburg Ltd, 1973.

'It is difficult to convince many breeders that it is possible to "over-egg the pudding" by concentrating too many influences for high quality in a pedigree. The result tends to be the production of too much nervous energy, producing over-excitability, nervousness or constitutional weakness.'

The incidence of stallions who were not themselves top class racehorses or stallions in the pedigrees of brilliant performers is remarkably high—for example Cecil in the pedigree of the Prix de l'Arc de Triomphe winner Rheingold and Caruso and Brown Bud in the pedigree of the American Triple Crown winner Secretariat. The Chismon adage may be interpreted as a striking and eloquent plea for balance in breeding plans, and for proper attention to all the factors required in a perfect racing machine.

*Two Aspects of the Racecourse Test*

The probability that the single-minded pursuit of speed at all costs may lead to errors of selection suggests a dual purpose for the racecourse test. Properly used, the test should designate the animals of both sexes that are most valuable for breeding not on grounds of speed alone but on grounds of speed and soundness combined. This is the point at which the doctrines of Persse and Lehndorff are reconciled. The two doctrines are not contradictory; they are the two sides of the same coin. By this criterion the most valuable horse for breeding is not the one who shows brilliance once or twice and then loses his form, or is unable to run again owing to some flaw of temperament, constitution or soundness. A horse of this kind should be viewed with reserve by breeders. The true stars are the horses who can show brilliance through a series of races and lose none of their form or physical condition. By this standard horses like Hyperion, Nearco and Ribot are supreme.

The question arises whether the rigour of the racecourse test can be overdone. It is argued sometimes that much hard racing may have a harmful effect on a horse's stud prospects and that a policy of restraint is advisable, especially where fillies and mares are concerned. The answer must be that no rule which is universally applicable can be stated. The amount of racing that can be sustained without injury varies with each individual. At the same time, it is important to bear in mind that the nature and combinations of the genes that are held in the sex cells and transmitted to the offspring will not be affected by the mildness or severity of the racing programme. The harmful effect of too much or too

*(17)* Mumtaz Mahal. Known as 'The Flying Filly', Mumtaz Mahal was brilliantly fast and has been a profound influence for speed in the modern thoroughbred. She was foaled in 1921

*(18)* Aloe. Foaled in 1926, Aloe became the ancestress of many important horses including Aureole, Parthia, Round Table and Alcide

*(19)* Hyperion. The winner of the Derby and St Leger in 1933, Hyperion became one of the foremost Classic sires in the mid-twentieth century

*(20)* Panorama. Foaled in 1936, Panorama was a magnificent physical specimen, and an outstanding sprinter and sire of sprinters

*(21)* So Blessed. Foaled in 1965, he was a top class sprinter. The prowess shown by him and his half-sister Lucasland was instrumental in securing the admission of their dam Lavant to the General Stud Book

*(22)* Nasrullah. Foaled in 1940, he won the Champion Stakes. He was leading sire of winners once in Great Britain and five times in the U.S.A., and was one of the most influential stallions of the twentieth century

*(23)* Crepello, Lester Piggott up. Foaled in 1954, he won the 2,000 Guineas and the Derby, and became a leading Classic stallion

(*24*) Nearco. Foaled in Italy in 1935, Nearco was the unbeaten winner of fourteen races and became a brilliant sire

hard racing are limited to the injuries to the reproductive organs and processes that may result from excessive physical strain and, in the case of a brood mare, to impairments of physique, nervous system and constitution which may produce an unfavourable environment for her foals before or after birth.

It is not difficult to find examples of mares who have had a great deal of racing and also achieved exceptional success at stud. Alice Hawthorn and Kincsem were two such mares. Alice Hawthorn, foaled in 1838, ran in seventy-one races, won fifty-one and dead-heated for another. The races she won included the Goodwood Cup, the Chester Cup and the Doncaster Cup twice, long-distance races of the most exhausting kind. As a brood mare she produced the Derby winner Thormanby, the Ascot Gold Vase winner Oulston and other good winners in Lord Fauconberg and Findon. Kincsem, the most celebrated thoroughbred product of Hungary, was foaled in 1874. She ran fifty-four times and was never beaten, racing in five countries and capturing, as her most important prizes, the Hungarian Oaks and St. Leger, the Austrian Derby, the Grand Prix de Deauville, the Goodwood Cup and the Grosser Preis von Baden (three times). Afterwards she bred one high-class offspring after another, one being the filly Budagyonge, who won the German Derby.

Even in cases where a hard racing programme has been followed by a disappointing stud career, it may be incorrect to ascribe the disappointing stud performance to an excess of hard racing. Pretty Polly, foaled in 1903, was one of the greatest racemares ever produced in England or Ireland. She won twenty-two of her twenty-four races, and her victories included the 1,000 Guineas, Oaks and St. Leger. Her stud record was not nearly so impressive, as no more than four of her offspring were winners and of these only the Cheveley Park Stakes winner Molly Desmond and the National Breeders Produce Stakes winner Polly Flinders were within hail of the top class. Not unnaturally, her relative lack of success at stud has been attributed to the debilitating effects of her exertions on the racecourse, but the truth was that she was too nervous to be a perfect matron. Noble Johnson, the stud manager to her owner and breeder Major Eustace Loder, made this note of her: 'Polly was not the sort of animal you could take liberties with, and if anything went wrong and she got upset, there was no knowing what might happen.' This nervous temperament was transmitted to most of her offspring. In addition, as a result of her own nervousness, she was not a good nurse to her foals, most of whom were delicate as a result.

There is little doubt that, if all the circumstances were known, opinions would have to be revised in many other cases in which failure at stud has been attributed to an excess of hard racing. In most cases the adverse effects of hard racing have probably been limited to temporary interference with normal reproductive processes. It may be significant that both Alice Hawthorn and Pretty Polly were barren for their first two stud seasons. After that Alice Hawthorn had a foal in each of the next seven seasons, and had ten foals altogether. Pretty Polly slipped twins in her third stud season, but produced a foal in each of the next ten seasons except one, in which she was not covered. Like Alice Hawthorn, she had ten foals altogether, a figure which is well above the average of about seven for the breed.

*Unsoundness*

Some common forms of unsoundness are respiratory troubles, blood-vessel breaking, heart trouble and various kinds of lameness which result from the inability of joints, bones, tendons and ligaments to stand the strain of prolonged galloping at racing pace. Many kinds of unsoundness are the consequence of faults of conformation, the term 'conformation' meaning the whole physical structure of the horse as a racing machine. These faults of conformation include straight shoulders, upright pasterns, and weak or misshapen hind legs.

Whenever a breeder uses an animal with an unsoundness or a serious fault of conformation for breeding he is taking the risk that the defect may be repeated in the offspring. A breeder in search of the ideal would reject all but perfect stock, but he would also need unlimited resources in order to apply this principle throughout his breeding operations. Most breeders have to compromise and take calculated risks, striking a balance between the defects and the strong points of the stallion and the mare selected for mating. This problem arises in its most acute form in inbreeding, as we saw in the case of Blandford; but in every mating, whether inbred or outbred, involving a horse of Blandford's type, the dangers and the possible advantages have to be weighed up.

Since the elimination of all animals showing any traces of unsoundness or faults of conformation is not practicable, the policy should be to use animals whose good qualities compensate for the defects, and to mate them with animals who are free from the same defects. If a mare has upright pasterns, she should be mated with a stallion who is flawless in this respect. If a mare has strong, straight and well-formed hind legs, it may be worth taking a chance by sending her to a stallion whose hind legs are less than perfect, provided that he is excellent in most other respects. Matings of this kind are often successful in correcting the fault in the offspring.

These 'corrective matings', as they are called, may do the trick as far as the immediate offspring are concerned. But it is important to remember that the offspring, though perfectly sound themselves, may carry the unsoundness as a recessive which is liable to be uncovered in some future generation. It is equally important to bear in mind that in order to correct a fault by selective mating the animal showing the fault should be mated with an individual who is flawless in that respect, not an animal showing a tendency to the opposite fault. The mating of extremes usually produces one extreme or the other, not a correction, in the progeny. Thus an animal with upright pasterns should be mated with one posesssing pasterns of perfect slope and length, not with one possessing pasterns so horizontal that the fetlock joints are practically touching the ground.

A fault, so far as it is heritable, is normally due to a recessive gene which must be dominated by the normal gene controlling the desired character, and cannot be averaged out with another recessive. Some kind of blending to eliminate a fault would be feasible only in cases where the character concerned is the result of the interaction of several genes.

### *The Arrangement of Individual Matings*

We have studied some of the general principles which must govern selection and indicated the necessity for a system of priorities which omits no vital consideration. The breeder still has many detailed questions to answer when he is planning each particular mating.

The practical problems of the breeder have to be solved in the light of his aims and the resources available. The aims of the commercial breeder, whose policy must be to satisfy the demands of the market, are not always identical with those of the private breeder, who has only himself to please; and the aims of the breeder with large financial resources may be more ambitious than those of the breeder who has to work on a limited budget. Nevertheless there are certain tests of suitability which have to be applied to both the partners in a mating, whatever the aims and resources of the breeder may be. In judging the compatibility of stallion and mare the following points are relevant in each case:

1. The horse's own racing ability and physical and temperamental characteristics.
2. The racing ability and characteristics of the horse's parents and other close ancestors.
3. The racing ability and characteristics of the horse's previous offspring.
4. The fertility of the horse.

Since the average quality of the stallions is bound to be higher than the average quality of the mares in the breed as a whole, the stallion chosen must be the best available to suit the purpose of the breeder. The best is not necessarily the most expensive stallion in the category concerned. Chamossaire stood at a fee of 300 guineas when he got Santa Claus, the winner of the Irish 2,000 guineas, the Derby and the Irish Sweeps Derby of 1964; Sayajirao stood at a fee of £250 when he got Indiana, the St. Leger winner of the same year; and Queen's Hussar also stood at a fee of £250 when he sired the brilliant and marvellously consistent Brigadier Gerard. In the year in which Brigadier Gerard was conceived, 1967, a nomination to Crepello, one of the most fashionable stallions of the day, changed hands for 2,600 guineas at public auction.

*Nicks*

A 'nick' is said to occur when two particular animals, or representatives of two particular thoroughbred strains, consistently produce offspring of higher quality when mated together than when mated otherwise. St. Frusquin and Glare provided an example of a nick between two individuals; to matings with St. Frusquin, Glare bred the 1,000 Guineas winner Flair; the Imperial Produce Stakes winner Vivid; and Lesbia, winner of the Imperial Produce, Champagne, Middle Park and Coronation Stakes and July Cup. These were of much higher quality than the offspring she bred to matings to other good stallions like Isinglass, St. Simon and Gallinule.

A famous nick between thoroughbred strains was between Phalaris and mares by Chaucer. This cross was responsible for the top-class horses Fairway, Pharos, Fair Isle, Caerleon, Colorado and Warden of the Marches. Precipitation got a high proportion of his best progeny when mated with mares descended from Phalaris in the male line; the notable products of this cross included Supreme Court, Premonition, Chamossaire and Sheshoon. This nick has been expressed in matings involving Sheshoon. Two of the best of the progeny of Sheshoon were the French Derby and Prix de l'Arc de Triomphe winner Sassafras, whose dam was by Ratification (grandson of Fair Trial): and the 2,000 Guineas winner Mon Fils, whose dam Now What was inbred to Fair Trial in the third generation. Fair Trial was a grandson of Phalaris.

The factors responsible for nicks are similar to those responsible for the phenomenon of hybrid vigour. The characters most commonly transmitted by one of the two strains are complementary to those most commonly transmitted by the other, and so matings involving a union of the two strains tend to produce a higher proportion of good offspring than other matings.

The Nasrullah – Princequillo nick provided the most striking illustration of the way this process may unfold. Nasrullah, imported into the United States after he had been leading sire of winners in Great Britain and Ireland once, became American leading sire of winners five times. He had been a brilliant racehorse, but had shown marked instability of temperament. Princequillo, imported into the United States as a foal, became the best stayer of his day and won the two most important American tests of stamina, the Jockey Club Gold Cup (2 miles) and the Saratoga Cup (1¾ miles). He was leading sire of winners twice. His stamina and equable temperament provided the perfect foil to the brilliance and instability of Nasrullah, and the Nasrullah – Princequillo cross has proved the most effective nick since the combination of Eclipse and Herod transformed the thoroughbred at the end of the 18th century. The paradigms of the Nasrullah – Princequillo cross have included the Derby winner Mill Reef, the Prix de l'Arc de Triomphe winner San San and the American Triple Crown winner Secretariat.

Bull Hancock, the master-breeder of Claiborne Farm in Kentucky where Nasrullah and Princequillo both stood, described nicks and their uses in these terms:*

'You like to drill where oil has been found. Nick may be a

* Quoted by Charles R. Koch in *The Blood-Horse*, February 1st, 1971.

bad word, but it gives you an outcross in which some things in the stallion compensate for their absence in the mare.'

There could be no clearer or more vividly phrased definition.

*Conclusion*

The modern breeder may be excused if he envies the simplicity of the problems that confronted the early breeders. In the seventeenth and eighteenth centuries, when the racehorse population was small and communications primitive, the breeder's choice of mates for his mares was practically limited to his own stallions and those of his neighbours. Moreover there was no diversity of aims since horses raced only as mature animals over long distances. As racing has developed so have the problems of the breeder multiplied and become more complex. Unless he is content to rely wholly on luck and illusion, the twentieth-century breeder has to renounce facile but untenable breeding theories, and instead make a realistic assessment of his aims and resources, strike a balance between the risks and the possible advantages of every proposed mating, and base his methods of selection on a comprehensive system of priorities.

# 8

## *Blood-lines and Families*

THE TERMS 'BLOOD-LINES' AND 'families' crop up repeatedly in discussions of the thoroughbred. Enough has been said already to indicate that too much weight has often been attached to them, but this aspect of breeding is so important that it is entitled to more detailed consideration.

In his book *Livestock Improvement* J. E. Nichols wrote: 'In a broad sense, all members of a breed which is descended from a relatively few ancestors must be related in some way, but as the common ancestors become more remote, their genetic influence upon their descendants must average less and less, and in fact is almost insignificant at three or four removes of generations. Only if there has been inbreeding or linebreeding, which keep genetic relationship high and increase the chances of similarity in genotype, can the genetic influence of an ancestor be considered of much account; such groups are best called "strains" or "lines". It is evident that a "line" in this sense differs in kind from the "blood-line" of thoroughbred usage, which refers to the unbroken chain of descent in the direct male or female line and exalts the influence of these "blood-lines", as far as the inheritance of characters is concerned, above the other component parts of a pedigree.'

Nichols proceeded to define the two kinds of group to which the term 'family' may properly be applied in animal breeding. The first is a group of close relatives consisting of parents, brothers and sisters, half-brothers and sisters, cousins and grandparents. The second is a group in which linebreeding, combined with selection, has been practised with sufficient intensity to separate the group from the rest of the breed. Nichols concluded: 'On the other hand, "family" has no special genetical significance as commonly used, when based only on a family name which may trace back either along the dams, or less frequently, through the sires.'

The ordinary human family is traced through the male line and identified by the family name. In many cases certain physical or mental peculiarities are attributed to human families more or less in perpetuity, and from time to time some convenient piece of evidence comes to light to support this supposition. For example, a seventeenth-century portrait exhibited publicly in London in the 1950s depicted a boy dressed in a white velvet suit, with blue sash, who bore an uncanny resemblance to an adult personality prominent in the racing world at the time of the exhibition. The catalogue revealed that the subject of the portrait was indeed the direct lineal ancestor of the twentieth-century racing personality. It is only too easy to leap to the conclusion, by a process of inductive reasoning, that physical and other characters are the property of human families in perpetuity. No doubt characters may be transmitted from generation to generation indefinitely as a result of dominance and gene linkage; but fact diverges from traditional beliefs about the inheritance of family characters because the transmission of these linked genes is not tied to the male line but may follow any of the many different routes that may be traced through a lengthy pedigree. Where the family name provides a check, a resemblance that spans a number of generations may be established readily in terms of heredity and hailed as proof of the potency of the direct male line. If the resemblance has not passed through the direct male line it is likely to escape notice, unless there is an observer possessing an intimate knowledge of the genealogical tree and of the family portraits.

The same principles apply to the thoroughbred, but in this case 'family' refers to the direct female line of descent—while 'blood-lines' may refer to the male or the female line. The work of Bruce Lowe and other researchers was responsible ultimately for this transposition, but the system adopted for recording the entries of mares and their produce in the General Stud Book renders it easier to trace the direct female line of descent than any other and therefore tends to emphasise 'family' in the Bruce Lowe sense.

The fact that all modern thoroughbreds may be traced in the male line to three foundation sires, the Byerley Turk, the Darley Arabian and the Godolphin Arabian—and to their connecting links in the second half of the eighteenth century, Herod, Eclipse and Matchem respectively—also has played a part in focussing attention on the female aspect of family. 'Families' traced in the male line would have no practical application, because those families would be too few and too vast to provide a useful method of differentiating various groups of thoroughbreds; especially as the Darley Arabian – Eclipse line has, in the evolution of the breed, grown to outnumber the other two lines, which have at times been in danger of extinction. There are enough surviving female lines, traceable from the earliest days of the racehorse, to make 'families' in this sense a basis of differentiating various groups, even though these groups may have little genetical significance.

As we saw in Chapter 4, the Bruce Lowe numbered families, based on the original foundation mares, have been largely replaced in modern usage by newer named families. The modern families are named after individual mares, mostly foaled within the last eighty years, who seem to have founded vigorous branches of the old lines or given them fresh life and impetus. These named families have never been classified as systematically as the Bruce Lowe families, but have become established in the phraseology of breeding. They have, it is true, more relevance to the breeding problems of the present day, but they are open to the same fundamental objection—that the attribution of qualities to the families as such is mistaken. On the other hand, there is a fairly widespread belief that some of these families derived from relatively modern 'tap-roots' possess special qualities that are carried forward from generation to generation; for example that the family of Black Cherry, foaled in 1892, has enduring resources of top-class, middle-distance ability and that the family of Americus Girl, foaled in 1905, is prepotent for top-class speed. Thus the essential question becomes: 'Can the contradiction between scientific fact, which is that prepotency is the property of individuals and not of families, and the apparent persistence of family characteristics over long periods, be resolved in any way?' The answer probably is that the continuance of a celebrated family is due neither to the undiminished influence of the tap-root mare nor to the inherent quality of the family as such, but should rightly be attributed to the careful selection of the best representatives of the family in each generation for breeding purposes, and an equally careful selection of their mates.

The part that intensive selection may play in the development of a family is illustrated by tracing the descent of Petite Etoile, one of the great racing mares of thoroughbred history, generation by generation from Americus Girl, her sixth dam. The family was called after Americus Girl because she marked the emergence of the female line from relative obscurity. Her dam Palotta was a second-rate performer with an undistinguished background, but Americus Girl herself, by the American-bred sprinter Americus, was fast enough to win the Fern Hill Stakes at Royal Ascot and the Portland Handicap. The connecting links between Americus Girl and Petite Etoile in the female line are Lady Josephine, Mumtaz Mahal, Mah Mahal, Mah Iran and Star of Iran—all mares who bore the hallmarks of judicious selection. Lady Josephine showed great speed as a two-year-old, when she won the Acorn and the Coventry Stakes; she was by the top-class sprinter Sundridge, who also sired the Derby winner Sunstar and the Oaks winner Jest. Mumtaz Mahal, who won seven races over short distances and was second in the 1,000 Guineas, was so fast that she was nicknamed 'The Flying Filly'; she was by The Tetrarch, one of the most brilliant horses in the history of the Turf. Mah Mahal won only two small races, but was by the wartime Triple Crown winner and great sire Gainsborough. She was a half-sister of Mirza II, one of the fastest two-year-olds of 1937.

The opportunities for Mah Iran to distinguish herself were restricted by the Second World War, but she was one of the fastest members of her age group and was allotted only four pounds less than Sun Chariot, the leader, in the two-year-old Free Handicap; she was by the Triple Crown winner Bahram and was a three-quarters sister of the Derby winner Mahmoud. Star of Iran won only one small race, but was by the Derby winner and high-class sire Bois Roussel; she was a sister of the Eclipse Stakes and Prix de l'Arc de Triomphe winner Migoli. Petite Etoile herself was by Petition, who showed exceptional speed as a two-year-old and won the Eclipse Stakes two years later.

To sum up—all the mares in the direct female line of Petite Etoile's pedigree as far back as Americus Girl were winners of some sort, and high-class winners in the majority of cases. Each was by a top-class racehorse and sire. Those of the mares in the chain of heredity who did not show exceptional racing ability themselves were closely related to outstanding performers. It is the consistently high standard of selection applied in each generation, not membership of the Americus Girl family as such, which is significant in the pedigree of Petite Etoile. The product of this process of intensive selection had the Free Handicap, 1,000 Guineas, Oaks, Champion Stakes, Sussex Stakes and Coronation Cup (twice) among her victories.

The objection is sure to be raised that the continuity of celebrated families is often assured by non-winning members. Aunt Clara, the dam, Sister Clara, the granddam, and Clarence, the great granddam of Santa Claus, did not win a race between them, so the victories of Santa Claus in the Derby and the Irish Sweeps Derby have been hailed as proof of the still-potent influence of Black Cherry, his sixth dam, which has miraculously reasserted itself after lying dormant for several generations. In the past the Black Cherry family has been responsible for horses like Blandford, Sun Chariot and the Oaks winner Carrozza and, so the argument runs, retains the capacity for throwing up a top-class middle-distance horse from time to time. The case for the persistence of family characters here seems strong, but analysis of the breeding of Santa Claus indicates that it may be more realistic to interpret his victories as evidence of the decisive importance of selection in the development of the thoroughbred.

Sister Clara, the granddam of Santa Claus, was by Scarlet Tiger, who was third in Hyperion's 1933 St. Leger but had only moderate success as a sire, out of Clarence. Sister Clara did not run and was covered for the first time as a three-year-old. Presumably there were no great expectations of her at the outset of her stud career, because her first three mates were Greek Bachelor, Stardust and Pactolus, of whom only Stardust, who was second in the 2,000 Guineas and the 1940 substitute St. Leger, could be considered of much account. The mating with Stardust produced Sarah Clara, the winner of the Stanmer Plate of £151 over five furlongs at Brighton as a two-year-old.

The picture underwent a dramatic change when the greatness of Sun Chariot, Sister Clara's younger half-sister by Hyperion, became apparent. As a close relative of a brilliant racing filly, Sister Clara was considered worthy of the best stallions available and was mated successively with Dastur, who had been second in all three colts' Classic races besides winning the Irish Derby and the Coronation Cup: with Hyperion, winner of the Derby and St. Leger and a vastly superior individual to his son Stardust; with another Derby winner, Straight Deal; with the

Eclipse Stakes and Prix de l'Arc de Triomphe winner Migoli; and with a third Derby winner, Arctic Prince. The mating with Arctic Prince produced Aunt Clara, the dam of Santa Claus. As a result of these matings, and further matings with the top-class sprinter and 2,000 Guineas runner-up The Cobbler, the stud record of Sister Clara showed marked improvement, and she bred seven winners of eleven races value £3,060 altogether.

Aunt Clara did not win a race, but her failure to do so was due not to lack of ability but to her headstrong temperament, which caused her to run herself into the ground well before the end of the three sprint races in which she competed as a three-year-old. She had plenty of speed. Provided that her flaw in temperament could be controlled by her matings, she obviously had the capacity for producing offspring of value. The St. Leger winner Chamossaire proved the right sire, and the mating with him produced Santa Claus. By the time of that mating Chamossaire had become unfashionable and was neglected by the majority of breeders trying to breed Classic animals, but he had shown his ability to get high-class stock by siring the St. Leger winner Cambremer and the Irish Derby winners Your Highness and Chamier. It was fashion that was at fault, not Chamossaire.

This account of the stud career of Sister Clara suggests that the importance of 'family' needs to be reassessed. As long as she was mated with undistinguished stallions she showed no sign of breeding anything but moderate offspring; but for her relationship to Sun Chariot she would probably never have been sent to top-class stallions, and she and her offspring would have remained in obscurity. Hyperion, not Clarence or membership of the Black Cherry family, may be presumed to have been the decisive factor in the excellence of Sun Chariot; and Arctic Prince, not Sister Clara or Clarence or membership of the Black Cherry family, may have been the decisive factor in the lower half of the pedigree of Santa Claus.

If belief in families in the Bruce Lowe sense is untenable, the question arises for how many generations a line derived from a prepotent tap-root may exert an influence worth considering. One answer, and an answer that is not entirely frivolous, is that a female line may have some value as long as breeders believe that it is valuable. Mares belonging to a highly esteemed family tend to be mated with good sires and thus the line receives infusions of the genes that are liable, sooner or later, to result in the breeding of a successful racehorse. When this happens the advocates of 'family' will hail it as proof of the stubborn survival of the admirable qualities of the line, which have been dormant for several generations, whereas what has really occurred is that, genetically speaking, a new family, owing little or nothing to the original tap-root, has emerged.

One of the dangers of obsession with 'family' is that it may lead to serious undervaluation of the other elements of a pedigree. Many unfashionable female lines are capable of rapid upgrading if subjected to the intensive selection that is applied habitually to the fashionable lines—that is, if the best and soundest members are mated to good sires. A dramatic example of the upgrading process is provided by the case of Prince Royal, the winner of the Prix de l'Arc de Triomphe in 1964. This top-class middle-distance horse had no glamorous family background as far as the dam's side of his pedigree is concerned. Penny Forfeit, his fifth dam, managed to win a selling handicap at Hurst Park after running ten times, an achievement which fairly reflected the status of the family at that period. Penny Flyer, his fourth dam, was a remarkable brood mare in her modest way, producing nine winners of 33½ races worth £8,167, though not one of those winners won a race worth more than £895. One of Penny Flyer's winning offspring was Prince Royal's great-granddam Princess Galahad, whose sire Prince Galahad won the Chesham and the Dewhurst Stakes as a two-year-old, but failed to train on. Princess Galahad won a selling race over five furlongs and dead-heated for a nursery over six furlongs at Wolverhampton as a two-year-old, and won a six furlongs handicap at Chester two years later—her winnings amounting to £431.

Princess Galahad was a second-class sprinter, and not one of the offspring of Penny Flyer was better than second class. On the other hand there was unmistakable evidence of merit in Princess Galahad and her half-brothers and sisters, and Princess Galahad confirmed this merit at stud by producing the Cambridgeshire winner Artist's Prince. Her daughter, Prince Royal's granddam York Gala, never ran, but had a good deal to recommend her in other respects. In the first place, she was by His Grace, a brother of the Derby winner Blenheim; in the second place, she bred seven winners, one of whom was the high-class handicapper Sterope, winner of the Cambridgeshire twice and of the Royal Hunt Cup. Another of the winners was Prince Royal's dam Pange, whose sire King's Bench was a top-class miler, winning the St. James's Palace Stakes and finishing second in the 2,000 Guineas. Without being anything out of the ordinary, Pange possessed useful speed and won races over five furlongs at Redcar and six furlongs at Manchester as a two-year-old.

The striking aspects of this family evolution are the ability of the tap-root mare to win, albeit in the lowest class, and the wide spread of respectable racing form in the next two generations. These were reasonably solid foundations on which to build, and the admixture of genes controlling high-class ability from His Grace, King's Bench and, finally, a great horse like Ribot, completed the upgrading process.

*Further Aspects of Family*

However limited the value of 'family' may be in the sense of the actual inheritance of characters, there is no doubt about the commercial value of 'family'. Membership of a fashionable family helps to sell foals, yearlings, fillies out of training and brood mares, and no commercial breeder can afford to disregard this factor. Nor is attachment to the idea of 'family' a matter of promoting a sales counter with many breeders, but represents a sincerely held belief. Nesbit Waddington wrote in an article in *The British Racehorse*:* 'There is, I know, a school of thought in thoroughbred breeding which maintains that a name as distant as even Pretty Polly's or Lady Josephine's can have no real influence on the progeny of today. I strongly disagree with this because I cannot see how so many good horses of today trace to such great family names if there is not some excellence in them which is transmitted to this day.' It is, as we have seen, not a question of some indestructible quality of a great mare that is passed down through an indefinite number of generations, but of the selection applied in each generation. On the other hand, it is possible to agree that the careful selection applied to the descendants of a great mare in each generation may result in the maintenance of a considerable degree of uniformity of genotype among many of those descendants. For this reason membership of a well-known family may provide a useful means of conveying the general pattern of a pedigree, provided that the individual under consideration is closely related to the main stream of the family's development. This aspect of the problem necessitates the brief account of some of the most

* Summer Issue, 1964.

celebrated named families of the present day which follows:

1. The Americus Girl Family. This is called alternatively the Lady Josephine family, after Americus Girl's daughter, but the wonderful speed which was the motive for applying intensive selection to the family was shown originally by Americus Girl, who was foaled in 1905 and should be regarded as the tap-root. The family chain has already been traced link by link from Americus Girl to Petite Etoile. The family has produced other top class middle distance horses like Mahmoud, Migoli, Nasrullah, Lady Juror and Commotion, and even one true stayer, the French St. Leger and Prix du Cadran winner Scot. Many more members of the family have possessed brilliant speed but limited stamina. These include Mumtaz Mahal, Tudor Minstrel, Abernant, Rivaz, Palariva, Combat, Diabler-etta, Mah Iran, Royal Charger, Tessa Gillian, Test Case, Gentle Art, Fair Trial, Mirza II, Kashmir II, Kalamoun and Habat.

2. The Black Cherry Family. Foaled in 1892, Black Cherry is the ancestress of the Classic winners Sun Chariot, Carrozza, Santa Claus, Night Hawk and Cherry Lass, and other top-class horses including Blandford, Blue Train, Pindari, Krakatao, Bebe Grande, Opaline II, Ragtime and Pieces of Eight. The manner in which the destiny of this family has been shaped decisively by selection has already been described.

3. The Pretty Polly Family. With three Classic and many other victories in important races to her credit, Pretty Polly, foaled in 1901, ranked as one of the greatest racemares of all time. Her descendants include top class horses of such widely varied talents as Donatello II, The Cobbler, Premonition, Supreme Court, Cappiello, Abadan, Only For Life, Double Bore, Arabella, Pardao, Arctic Explorer, Thymus, Vienna, Great Nephew, Luthier, Daring Display, Don II, Felicio, Huntercombe, Lucyrowe, Favorita, Cavo Doro and Brigadier Gerard. The family was represented by two successive Derby winners, St. Paddy and Psidium, in 1960 and 1961. But enthusiasm for this evidence of enduring 'family' influence must be tempered by the reflection that St. Paddy and Psidium traced their descent from different daughters of Pretty Polly, Molly Desmond and Dutch Mary respectively, and that Pretty Polly herself was six generations away from St. Paddy and seven generations away from Psidium. Brigadier Gerard, the family's most distinguished member although he did not run in the Derby, traced his descent from Molly Desmond.

The Pretty Polly family was named alternatively after Pretty Polly's dam Admiration in order to embrace another varied assortment of talents including Tehran and Gratitude in the family net. The attempt to find a genetical common factor between, say, Gratitude and Psidium, must involve stretching the conception of 'family' to breaking point.

4. The Sceptre Family. Foaled two years before Pretty Polly, Sceptre exceeded Pretty Polly's Classic score by one, failing in only the Derby, in which she was fourth, of the 1902 Classic races. The Sceptre family had its first Derby winner, Relko, two years after the Pretty Polly family had its second. Relko was separated from Sceptre by seven generations. Two years later again Reliance II gave the family its first victory in the French Derby. Match III had won the French St. Leger in 1961. Relko, Reliance II and Match III won seven English and French Classic races between them. They were all out of the same mare, Relance, and it may be argued that the complementary influences of Relance's American sire Relic and Tantieme (sire of Match III and Reliance II) and Tantieme's son Tanerko (sire of Relko) counted for a good deal more than membership of the Sceptre family. Other top class horses descended from Sceptre include Northern Light, Midnight Sun, Craig an Eran, Sunny Jane, Tiberius, Buchan, St. Germans, Noor, Zucchero, Flyon, Petition, Torbido, Tissot, Tamanar, Taboun and Full Dress II.

5. The Marchetta Family. This is another well-known family which was represented by its first Derby winner in the early 1960s. This was Larkspur, who was successful in the trouble-stricken Derby of 1962, in which seven horses fell. Nine years later the family had its first Oaks victory with Altesse Royale, who had previously won the 1,000 Guineas. Larkspur and Altesse Royale both traced their descent from Marchetta's daughter Rose Red. The Rose Red branch has produced many other top class horses including the great stayer Alycidon, Borealis, Even Star, Acropolis, Festoon and Celtic Ash.

The other main branch of the family had its origin in Marchetta's daughter Sweet Lavender. This branch has produced top class horses like My Babu, Sayani, Ambiorix, Cagire II, Turn-To, Klairon, Haltilala, Dan Kano and English Prince.

1974 was a remarkable year for the Marchetta family, for Bustino won the St. Leger, English Prince won the Irish Sweeps Derby, and Imperial Prince was second in the Derby and the Irish Sweeps Derby. Marchetta was foaled in 1907.

6. The Frizette Family. This family has achieved unusually wide geographical spread, as descendants of Frizette, who was foaled in the U.S.A. in 1905, have distinguished themselves in many countries, particularly the U.S.A., France and England. Frizette was extremely prolific, and no fewer than seven of her daughters—Banshee, Frizeur, Lespedeza, Durzetta, Princess Palatine, Ondulation and Frizelle—founded branches of the family which thrived. The family has had two English Classic winners, the 1948 St. Leger winner Black Tarquin and the 1964 2,000 Guineas winner Baldric II. They were both descendants of Princess Palatine.

The Banshee and Frizelle branches provided some of the greatest successes of the Marcel Boussac stud in France. The Banshee branch provided the French Derby winner and prepotent sire Tourbillon, Priam II, Djeddah, Cordova, Corejada, Apollonia and Macip. The Frizelle branch provided the French Derby winners Cillas and Auriban.

Frizeur was the ancestress of the Kentucky Derby winner Jet Pilot and the excellent American mare Typecast, and Ondulation was the ancestress of Dahlia, winner of the Prix Saint Alary, the Irish Guinness Oaks, the King George VI and Queen Elizabeth Stakes and the Washington D.C. International in 1973.

In the 1970s the genetic similarity between members of the different branches, separated by six or more generations from the tap-root, must be slight in most cases.

7. The La Troienne Family. Like Frizette, La Troienne was very prolific. Foaled in France in 1926, she spent her whole stud career in the U.S.A. She bred ten winners including the Classic winners Bimelech, successful in the Preakness and the Belmont Stakes, and Black Helen, successful in the Coaching Club American Oaks. Black Helen and eight other daughters of La Troienne—Baby League, Big Hurry, Big Event, Businesslike, Besieged, Bee Ann Mac, Belle Histoire and Belle of Troy—founded enduring branches of the family. The Baby League branch has provided the only English Classic winner, Boucher, successful in the St. Leger in 1972. This branch has provided many top class American horses including Busher, Harmonising, Hitting Away, Dapper Dan and Numbered Account.

Big Hurry was the dam of Searching, a brilliant racemare and dam of two mares of similar quality, Affectionately and Priceless Gem. Affectionately was the dam of Personality, winner of the Preakness Stakes in 1970, and Priceless Gem was dam of Allez France, winner of the French 1,000 Guineas, the French Oaks and the Prix Vermeille in 1973. Businesslike was the granddam of Buckpasser, winner of 25 races and nearly $1\frac{1}{2}$ million dollars. Other top class horses from the family include How Now, Cohoes, The Axe II, Autobiography and Malicious.

There may be considerable genetic similarity between different members of the La Troienne family. She is close up in the pedigrees of many of her descendants in the 1970s, and the fact that the American Triple Crown winner War Admiral also figures in the majority of the pedigrees has helped to preserve a marked degree of consanguinity.

8. The Aloe Family. Although she did not win a race Aloe, foaled like La Troienne in 1926, was by no means a negligible performer on the racecourse. The evolution of the family has depended on the success of her daughters Feola and Sweet Aloe. Feola herself was second in the 1,000 Guineas and third in the Oaks. Her descendants in England include the 1,000 Guineas winner Hypericum, the 1,000 Guineas and French Oaks winner Highclere, Angelola, Aureole, Kingstone, Above Board, Above Suspicion, Doutelle and Cadmus. This branch has also flourished abroad, and is responsible for the Italian St. Leger winner Ben Marshall, the Argentine Oaks winner Sideral and, in the U.S.A., for the top class racehorse and champion sire Round Table and the Hollywood Derby winner Tell. The French-bred Lassalle, a great granddaughter of Feola, won the Prix du Cadran and the Ascot Gold Cup in 1973. Sweet Aloe became the granddam of the St. Leger winner Alcide and the great granddam of the Derby winner Parthia.

9. The Simon's Shoes Family. Simon's Shoes was foaled in 1914, but it was not until the Second World War that her family began to make its mark indelibly in top class racing. The reason for this delayed success was that she bred the daughters who were destined to give the family fortunes their vital impetus, Carpet Slipper and Dalmary, when she was 16 and 17 years old.

Carpet Slipper's daughter Godiva won the 1,000 Guineas and the Oaks in 1940, and two years later her son Windsor Slipper won the Irish Triple Crown. Since then this branch of the family has produced a stream of top class winners including Val de Loir, Valoris, Reform, Lalika, Vincennes, Scherzo, Alciglide and Allangrange. The evolution of this branch had a remarkable climax in 1973 when Roi Lear, by Reform out of Carpet

Slipper's great granddaughter Kalila, won the French Derby.

Dalmary, herself a winner of the Yorkshire Oaks, founded a branch from which such good European winners as Lorenzaccio, Sammy Davis, Super Sam, Le Sage and the Irish St. Leger winners Barclay and Christmas Island have sprung. Her great grandson Tudor Era showed excellent form in the U.S.A. But it was the move of her daughter Rough Shod to the U.S.A. which set the family fortunes alight in that country. Rough Shod bred Moccasin, a filly so brilliant that she was 'Horse of the Year' at two years of age, and two other high class performers in Ridan and Lt. Stevens. This American branch has proved a prolific source of talented horses like Gamely, Drumtop, Thatch, Cellini and Apalachee.

Members of the Carpet Slipper and Dalmary branches had no more than a faint genetic relationship by the 1970s, and for practical purposes could be regarded as two independent families.

10. The Orlass Family. Orlass, foaled in 1914, won five races in Ireland, all over five furlongs. Many of her descendants have shown excellence over middle distances, which suggests that the stamina contributed by her own mates and the mates of mares in succeeding generations has played a vital part in the evolution of the family. The Orlass family has provided the English Classic winners Humble Duty (1,000 Guineas), Neasham Belle (Oaks), and Hethersett and Peleid (St. Leger). The numerous other top class performers that have embellished the family record in England include Narrator, Netherton Maid, None Nicer, Orestes, Gay Time, Elopement, Crisper, Cursorial, Nucleus, Derring-Do, Jacinth and Coup de Feu. Homeric showed good form in England and France, and the French 1,000 Guineas winner Dumka, Saraca, Simbir and Sybarite all shone in France. The Orlass family has also penetrated the United States, where its most successful representative Droll Role won the Washington D.C. International in 1972.

Very few members of the Orlass family have failed to stay at least a mile, and the large majority have been middle distance performers. This consistency of aptitude has made the family one of the most dependable sources of Classic material available to breeders.

All these ten family names are in common use when pedigrees are under discussion, though it appears that many of these named families must embrace assortments of individuals whose distant relationship can have little relevance to the real problems of breeding. In these cases 'family' does not even have the justification of describing a general pattern of breeding.

These ten families supplied top-class horses with considerable regularity for a quarter of a century and more after the Second World War. No attempt has been made to arrange them in any order of precedence, since the degree of prominence attained by each family varies from year to year. But it should be a sobering reflection for anyone prone to obsession with the importance of 'family' that ten others could be added to the families here described without embracing the female lines of Ribot, Crepello, Ballymoss, Right Royal V, Sea Bird II, Sir Ivor, Nijinsky, and Secretariat, who on any rational assessment must be reckoned among the best horses of the same period. A combination as potent as the Nasrullah – Princequillo nick may transcend all family influences. Evidently top-class racing ability is not the monopoly of a small number of families, and the danger of pre-occupation with this aspect of any pedigree is that it may divert attention from the possibly decisive merits or flaws contained in the rest of the pedigree.

*Male Lines*

Male lines may be developed, through selection, in exactly the same way as the families traced through the female line, so that special qualities come to be regarded as their property. In this way the Orby line became established as the principal pure sprinting line during the first half of the twentieth century.

Orby, by the Eclipse and Champion Stakes winner Orme out of the American mare Rhoda B, won the 1907 Derby, but was not regarded as a stayer by contemporary observers, though he had great speed. In his turn he sired another Derby winner, Grand Parade, but Grand Parade was neither typical of the progeny of Orby nor able to get a son to continue an Orby middle-distance male line. Most of the progeny of Orby were wanting in stamina to such an extent that their average winning distance above two years of age was little more than six furlongs.

One of Orby's speedy sons was The Boss, a successful sprint handicapper who gained his most important victory in the Gosforth Park Cup at Newcastle. Although largely ignored during his first few seasons at stud, The Boss forced himself upon the attention of breeders by sheer merit and got Sir Cosmo, Golden Boss, Heverswood and Scherzo, all very fast runners.

The success of The Boss at stud marked the decisive stage in the development of the Orby line, because it was considered, from then on, to be a source of speed pure and simple and breeders refrained almost entirely from sending any but speedy mares to sires of the Orby line. If Orby and his progeny tended to possess speed in excess of stamina, then the stud career of The Boss determined that the balance of staying power was unlikely to be shifted for some generations to come.

The Orby line divided into two branches after The Boss and was continued by The Boss's sons Sir Cosmo and Golden Boss. The Sir Cosmo branch was extended mainly by Panorama and Panorama's son Whistler, and the Golden Boss branch by Gold Bridge and Gold Bridge's sons Denturius, Golden Cloud and Vilmorin, all high-class sprinters and successful sires. Intensive selection in each generation has achieved a considerable degree of uniformity in the genotypes of the members of both branches, and the reputation of the Orby line as a source of precociously fast horses and sprinters has been consolidated firmly.

The breeding of Panorama and Gold Bridge, two vital factors in the process, illustrates the kind of selection that was responsible for this consolidation. Panorama, by Sir Cosmo, was out of Happy Climax, who won three and was placed second in four other races over short distances as a two-year-old, and also finished first on two occasions only to suffer disqualification. One of the races in which she was disqualified was the important Acorn Stakes at Epsom. Happy Warrior, the sire of Happy Climax, was another speedy and precocious performer, and won the Hopeful Stakes at Newmarket and two nurseries as a two-year-old. Although he was second in the St. James's Palace Stakes over a mile at Royal Ascot the next year, he was a non-stayer and afterwards reverted to sprinting. He was by Sundridge, who, in spite of being wrong in the wind, was exceptionally fast and won sixteen races over sprint distances between three and six years of age.

While the significance of Panorama's breeding lay in the accumulation of non-stayers close up in his pedigree, the salient point in the case of Gold Bridge was inbreeding to Orby, who was the grandsire both of his sire Golden Boss and his dam Flying Diadem. Indeed the breeding pattern involved in the production of Gold Bridge approached more closely to J. E. Nichols's definition of a 'line' than that of any other prominent member of the so-called Orby male line. It is not surprising that the average winning distance of the progeny of Panorama and Gold Bridge was less than six furlongs in each case.

At the stage of specialisation in speed represented by Panorama and Gold Bridge, and the majority of their descendants, the question is bound to arise: 'Is the Orby line destined to remain a pure sprinting line for ever, or at least as long as it is able to survive at all in the competitive world of racehorse breeding?' The answer must be that the prepotency for speed possessed by these two sires, and prepotency for any other character, cannot be the property of a male line any more than it can be the property of a 'family'. Prepotency is the property of an individual by virtue of holding the dominant genes concerned in double strength, and so is only one generation deep. Fresh methods of selection may bring about changes in the kind of genotype that has been regarded as typical of the line from one generation to the next. Gratitude (by Golden Cloud) and Quorum (by Vilmorin) were two stallions from the Golden Boss branch of the Orby line who sired horses capable of staying further than sprinting distances.

Golden Cloud and Vilmorin were typical of the line in the sense that they were pure sprinters themselves and got progeny whose average winning distance was only about six and a half furlongs. On the other hand neither Gratitude nor Quorum was out of a mare typical of the mates of sires of the Orby line. Verdura, the dam of Gratitude, also bred the Cesarewitch winner Avon's Pride, and had as the four sires in the third generation of her pedigree the great Classic sire Blandford and the St. Leger winners Fairway, Salmon Trout and Hurry On. Gratitude himself succeeded only as a sprinter, but must have inherited some of the genes for increased stamina and passed these on to many of his progeny, who included the Lincolnshire Handicap winner Mighty Gurkha and Solitude, winner of the valuable Prix de Flore over 1 mile 2½ furlongs at Saint Cloud.

Quorum stayed better than Gratitude and proved himself a top-class miler, finishing second to Crepello in the 2,000 Guineas and winning the Sussex Stakes at Goodwood. The sources of stamina close up in the dam's side of his pedigree were his grandsire Bois Roussel, winner of the Derby, and his great-grandsire Noble Star, winner of the Ascot Stakes, Goodwood Stakes and Cesarewitch. Quorum proved capable of siring genuine stayers—horses like Beddard, who won the Queen's Vase (2 miles), Quartette, who won the Yorkshire Cup

($1\frac{3}{4}$ miles) and Cullen, who won the Great Metropolitan Handicap ($2\frac{1}{4}$ miles), besides the Grand National winner Red Rum.

Gratitude and Quorum have indicated the possibilities of change, but it would be a gross exaggeration to suggest that further progress in the development of stamina is inevitable or even likely. A number of conditions would have to be fulfilled for any male line descendant of Orby to attain the prominence as a middle-distance sire that many past and present members of the line have attained as sprinting sires. One condition is a matter of chance, because every evolutionary step forward depends on the right combination of genes coming together at the right moment; another, and perhaps the decisive condition, is that sires of the Orby line should have mates of a quality, collectively speaking, to enable them to complete on equal terms with established middle-distance sires—in other words, that there should be a deliberate effort to deploy the forces of selection in their cause.

Whatever the future of the Orby line may be, Gratitude and Quorum are entitled to a high award for their help in destroying the myths that have surrounded the subject of 'blood-lines and families'.

# 9

# *The Stud Book*

THE GENERAL STUD BOOK AND THE annual volumes of *Races Past* are the twin pillars which support the evolution of the British Thoroughbred. The one provides the pedigrees of brood mares and their foaling records; the other provides the evidence of the racecourse test upon which the selection of the most suitable animals, male and female, for breeding purposes must be based. Without them no intelligent breeding policy would be possible.

There were, of course, annual 'Calendars' of racing results for half a century before James Weatherby published his first *Racing Calendar* in 1773, and in more recent times the official records have been supplemented by a wide variety of form books and commentaries published independently of the racing authorities. There were private stud books kept by many breeders long before a second James Weatherby, a nephew of the first, published *An Introduction to the General Stud Book* in 1791 and the first volume of the General Stud Book two years later. The word 'general' was used in the title of his work to differentiate it from the private records of their own bloodstock kept by individual breeders, mostly in haphazard fashion. In more recent times the General Stud Book has been supplemented by stallion registers, pedigree charts and numerous other independent publications. But the two original publications of the Weatherbys, which later received official sanction, have remained the basic and indispensable records, and the breeders of all subsequent periods, and the Turf in general, are eternally in debt to their creator.

We have seen that the rapid improvement of the racehorse

was just one aspect of the farming revolution, which included the introduction of scientific stock breeding, which took place in England during the eighteenth century. Bakewell, the pioneer of scientific stockbreeding in sheep, cattle and cart-horses, founded the Dishley Society to ensure purity of breed. This was the intellectual background of Weatherby's General Stud Book, but Weatherby made his own contribution to the revolutionary process, and put racehorse breeders even more deeply in his debt, by publishing his first volume nearly fifty years before the appearance of the first of the cattle-breed herd books.

The purpose of Weatherby in publishing his General Stud Book was, as he stated in the Introduction, to correct the 'increasing evil of false and inaccurate pedigrees'. Clearly the existence of a comprehensive register of pedigrees and foalings was calculated to reduce the risks of fraud and of genuine cases of mistaken identity, and Weatherby, apparently, was less concerned with the use of his stud book as a means to the development of the breed of the racehorse. Nor did he consider it necessary to define the qualifications for entry in the book, claiming merely that it contained 'a greater mass of authentic information respecting the Pedigress of Horses, than has ever before been collected together'. His sources were old racing calendars and sale papers, and information supplied by the principal breeders and other racing men. Evidently he set himself the task of collecting information and arranging it in orderly fashion, not of laying down rules of admission to a kind of exclusive club.

Nevertheless it would be incorrect to assume that Weatherby had no criterion of eligibility for the General Stud Book other than distinction as a racehorse or relationship to racehorses of distinction. He did not use the word 'thoroughbred' in Volume I, but implied its existence by using its antithesis, 'half-bred', to describe animals that could not be admitted. For example, he made a note concerning the Young Marske mare bred by Sir John Webb: 'In 1806 missed to Sir Peter, having previously bred 14 foals; those unnoticed here'—numbering three—'were by half-bred stallions.' The reverse of 'half-bred' could only be 'thoroughbred'.

The first mention of 'thoroughbred' in the General Stud Book was in Volume II, published in 1821, but without definition. Copenhagen, whom we have encountered in an earlier chapter, was included in that volume, together with the comment that the Duke of Wellington had ridden him at the Battle of Waterloo. He was by Meteor, and the pedigree of his dam Lady Catherine was given as: 'Got by John Bull, her dam by the Rutland Arabian, out of a hunting mare not thoroughbred.'

What was meant by 'thoroughbred'? Much later, in the middle of the nineteenth century, the *Encyclopedia of Rural Sports* observed unhelpfully: 'The term thoroughbred, as relating to a horse, is neither critically nor conventionally definite.' Fortunately this opinion was not held universally, for William Osmer, the veterinary surgeon and shoeing smith of Blenheim Street, had stated in his *Treatise on the Horse* as early as 1761: 'Let us suppose a case: here are two mares, both originally bred from Arabian horses, and mares, or the descendants of such, which I suppose is all that is to be understood by the term "thoroughbred".' Probably Weatherby adopted the same standard to differentiate between thoroughbred and half-bred; and this probability obtained some corroboration from the pen of a later editor, who stated in the Advertisement to Volume XIV of the General Stud Book: '. . . a recent importation of Arabians from the believed best Desert strains will, it is hoped, when the increase of size has been gained by training, feeding and acclimatisation, give a valuable new line of blood from the original source of the English thoroughbred.'

If thoroughbreds were horses descended, or believed to be descended, from Eastern stock, and half-breds were ineligible for the General Stud Book, then the example of Copenhagen signifies that the editors of the earlier volumes did not apply this principle with perfect consistency. Copenhagen, after all, was a winner, and the decision to admit him may have been swayed by his historical interest or by deference to the national hero who owned him.

These two aspects of the policy of the editors of the early volumes of the General Stud Book—an implicit qualification for admission combined with flexibility in applying the principle—were to have important consequences for the development of the British racehorse about a century after the publication of James Weatherby's original volume.

*Events leading up to the Jersey Act*

British racehorses and breeding stock began to be exported before the end of the eighteenth century. The Preface to Volume II of the General Stud Book noted that attempts were being made in a number of foreign countries to establish breeds of racehorses on the English plan, together with a more careful selection of stallions and brood mares than was the practice in England. Foreign breeders were confident that their improved methods would enable them soon to produce better horses than the English, who would be compelled to send abroad for sound, if not for speedy, horses. The Stud Book editor commented: 'This hint about soundness may be worth attention, but for the rest, with the advantages this country already possesses, and so long as horse-racing continues to be followed up with spirit by her men of rank and opulence, there can be little to apprehend.'

This complacency was pricked when foreign horses began to win important English races after the middle of the century, and English breeders began to import foreign bloodstock as a consequence. The Stud Book editors did not refer to England's natural advantages again, but the Advertisement to Volume XIII of the General Stud Book—the Stud Book had been published at four-yearly intervals since the publication of Volume VI in 1849—in 1877 contained this paragraph: 'It will be noticed that a considerable number of mares and stallions now in this country were bred abroad, especially in France; and in recording the pedigrees, etc., of these animals, the Publishers have been much assisted by the excellent Stud Books now published in France, Germany, Austria, the United States of America, and the Australian Colonies, and also by the certificates of pedigrees required by the Rules of racing to be produced before foreign bred horses can run in this country. Only those foreign bred horses are included in this work, which have been or are likely to be used for breeding purposes.'

At this stage the editors apparently felt no twinge of concern, but as the American and Australian breeding industries continued to expand rapidly and more and more of their products began to come to England, the Stud Book authorities were forced to take a closer look at the consequences. By the time of the publication of Volume XVIII twenty years later all complacency had vanished and been replaced by a faintly querulous note in the Advertisement, which contained the sentence: 'The importation of a number of horses and mares bred in the United States of America and in Australia, a few of which will remain at Studs in this country, may have some effect on stock here, but the pedigrees of these horses, though accepted in the Stud Books of their own country, cannot in all cases be traced back to the thorough-bred stock exported from England from which they all claim to be, and from which, no doubt, they mainly are descended; these animals are, therefore, in these cases, marked with reference to their own Stud Books.' Further consideration revealed that this method of treating the difficulty could lead only to confusion. Although the General Stud Book was still the property of the Weatherbys, the editors took the unprecedented step of seeking the advice of the Stewards of the Jockey Club, as the highest authority on all matters connected with the Turf. Their deliberations produced the first qualifying test for admission to the General Stud Book, to which all imported horses and mares included in Volume XIX had been submitted. The qualification was that 'any animal claiming admission should be able to prove satisfactorily some eight or nine crosses of pure blood, to trace back for at least a century, and to shew such performances of its immediate family on the Turf as to warrant the belief in the purity of its blood'.

This restriction was to prove an inadequate safeguard in the face of events overseas. American racing was undergoing a crisis in the first decade of the twentieth century, and at the end of that period anti-betting legislation practically brought the sport to a standstill in the United States for a number of years. The bottom fell out of the market for racehorses in that country, and the racing authorities in England had nightmarish visions of American animals of dubious pedigree being deposited in England by the shipload. These fears were responsible for the prefatory notice in Volume XXI of 1909: 'The Editors beg to inform Subscribers that, since the last volume of the Stud Book was published, they have had cause to consider the advisability of admitting into the Stud Book horses and mares which cannot be traced to a thorough-bred root, but which have fulfilled the requirements given in the preface to Volume XIX. They have decided that, in the interests of the English Stud Book, no horse or mare can be admitted unless it can be traced to a strain already accepted in the earlier volumes of the Book.'

This edict was intended to reinforce the barriers previously erected, but its terms were imprecise and ambiguous, and failed to appease protectionist sentiment in England. At the Jockey Club Meeting in the spring of 1913 Lord Jersey, who had been Senior Steward the previous year, put the case for making the regulations for admission to the General Stud Book still more stringent and, on the acceptance of his proposals by the Stewards and members, the editors inserted the following new rule for qualification in Volume XXII, published the same year: 'No horse or mare can, after this date, be considered as eligible for admission unless it can be traced without flaw on both sire's and dam's side of its pedigree to horses and mares themselves already accepted in the earlier volumes of the Book.'

This was the so-called 'Jersey Act'; it generated feelings of bitterness in many foreign countries, particularly the United States, which were not to be allayed until the repeal of the 'Act' thirty-six years later.

### *The Motives for the Jersey Act*

Americans saw the Jersey Act as a measure intended to deny them an outlet in world markets for their vast numbers of surplus animals—and to protect the British export trade in thoroughbreds at the same time—by putting the stamp of 'half-bred' on the majority of American horses. Few American horses escaped this stigma, mainly because the great sire Lexington appeared in the pedigrees of most horses bred in the United States by the turn of the century. Although Lexington traced back to Diomed in the male line, he had doubtful elements in the dam's side of his pedigree which disqualified him from the General Stud Book under the regulations of 1913. Curiously enough, Lexington himself had been the subject of a brief note in Volume X, published in 1865, under the heading 'American Horses'. This read: 'Lexington, a bay horse, foaled in 1850, by Boston, out of Umpire's dam.' This did not legitimise him under the terms of the Jersey Act.

It would be disingenuous to pretend that there was not a powerful lobby of interested parties in England. Indeed a majority of the Jockey Club members who accepted the Jersey Act and recommended it to the Stud Book authorities were breeders with an interest in protecting export prices as a means of disposing of their surplus stock. On the other hand, the most vocal foreign critics were unjust in dismissing as mere hypocrisy the British claim that the purpose of the Jersey Act was to defend the purity of the General Stud Book.

We have seen that the concept of 'pure blood' had existed in the minds of the Stud Book editors from the earliest days, although the principle had not always been applied consistently. Indeed the Preface to the first edition of Volume III, published in 1827, had contained an admission of possible accidental error in the statement that 'the English Stud Book, although so ably commenced, was begun too late, as appears from the imperfect accounts of many of the old stallions and mares in the first Volume.' Nevertheless the General Stud Book, once compiled and launched, was kept with scrupulous regard for accuracy. The attitude of the editors was bound to harden into exclusiveness if growing numbers of foreign-bred animals of doubtful pedigree began to claim admission. The story of this hardening process was told in the Advertisements and Notices to the Stud Books from Volumes XVIII to XXII.

Nor was the 'impurity' of the American Stud Book a myth invented solely by unscrupulous British protectionists. Doubts about the reliability of the work of Colonel Bruce, the compiler of the American Stud Book, were also expressed in some quarters in the United States. In an article published in the *Thoroughbred Record*, of Lexington, Kentucky, in March 1927, and reproduced in *The Bloodstock Breeders Review* the following year, John Hervey wrote of Bruce: 'In compiling the Stud Book, in giving to our official genealogies the forms into which they have crystallised as authentic, he did, in various, if not numerous instances, high-handed and unwarranted things which the facts, the evidence and the testimony extant, stamp as indefensible.' According to Hervey, Bruce was dependent on the patronage of leading American breeders in order to publish his work, and this dependence disposed him to give uncritical acceptance to the pedigrees offered to him. As a result, '. . . he shovelled into the Stud Book pedigrees so impossible as to be nothing less than absurd, and this gave to outrageous fiction the official stamp of veracity'. Hervey added that to 'trained scrutiny' the omissions, errors and contradictions were 'so obtrusive as to become both exasperating and inexcusable'.

When flaws were condemned so roundly in America—and it is scarcely to be imagined that Hervey was the first man to be aware of them—the English Stud Book authorities could not be blamed for being on their guard. The motives that inspired the Jersey Act may have been mixed, but they were not wholly dishonourable.

### *The British Thoroughbred after the Jersey Act*

Although all animals claiming admission to the General Stud Book after 1913 were submitted to the test of the Jersey Act, the test was not made retrospective. Thus American and other strains of dubious origin that were already in stayed in, and were as thoroughbred, if admitted only in 1900, as if they traced back without flaw to a mare who lived in England two hundred years earlier. Without this concession, the Jersey Act could hardly have remained in force as long as it did.

Three American horses imported into England in the 1890s had a profound influence on the development of the British Thoroughbred. These were Americus, imported as a three-year-old in 1895; Rhoda B, imported as a yearling in 1896; and Sibola, imported as a yearling in 1897. Americus, who had raced as Rey del Caredes in his native country, was a high-class sprinter and sired Americus Girl, whose descendants, from Mumtaz Mahal to Petite Etoile, include some of the most brilliant horses ever to have run in England. Rhoda B became the dam of Orby, winner of the Derby and a notable source of speed. Sibola won the 1,000 Guineas and was second in the Oaks, and became the great-granddam of Nearco. Americus, Rhoda B and Sibola were admitted to the General Stud Book; yet they all had Lexington in their pedigrees and would have been excluded if they had claimed admission after 1913. Indeed Americus was inbred to Lexington, his pedigree being given in Volume XIX as 'by Emperor of Norfolk (son of Norfolk, by Lexington), his dam, Clara D, by imp Glenelg, out of The Nun, by Lexington'.

The first serious blow to the integrity of the Jersey Act came almost at once, in 1914, when Durbar II won the Derby. Durbar II was by the British-bred Rabelais (by St. Simon), who won the Goodwood Cup and was third in the 2,000 Guineas and fourth in the Derby, out of Armenia, by Meddler out of Urania, a marvellously tough mare who won 35 of her 87 races. It was Urania who disqualified him from the General Stud Book, because she was by Hanover, whose sire, the Kentucky Derby winner Hindoo, was out of Florence, by Lexington. Rhoda B also was by Hanover. Thus the paradoxical situation had been created in which the Derby winner of 1907, Orby, was acceptable for the General Stud Book in spite of being out of a Hanover mare, but Durbar II, the Derby winner of seven years later, was debarred because he was the grandson of a Hanover mare. The exclusion of Durbar II had a deep significance for the future of the British Thoroughbred which was not fully appreciated at the time.

The contradictions soon began to multiply. In *The History of the Racing Calendar and Stud Book* C. M. Prior pointed out that

Hanover appeared twice in the pedigree of Diapason, the Goodwood Stakes winner of 1924, once in what may be called his legitimate and once in his illegitimate guise. Diapason was by Diadumenos, by Orby, who represented an accepted strain of Hanover. On the other hand Diapason's maternal grandsire Sir Martin who, like Orby, was out of a Hanover mare, was imported from America after the imposition of the Jersey Act and was excluded from the General Stud Book on account of his Hanover connection. Diapason's dam Venturesome and Diapason himself were both ineligible.

Despite the case of Diapason and others like it, the Jersey Act was upheld firmly until the Second World War. In the face of occasional victories of so-called half-bred horses in important races, the advocates of the Jersey Act asserted that it was 'purity of blood', not racecourse performance, that mattered. But the last colts Classic race to be run before the fall of France in 1940 presaged the ultimate collapse of exclusiveness. The 2,000 Guineas that year was won by the French-bred and trained Djebel, who was not eligible for the General Stud Book because his sire Tourbillon was out of a mare by Durbar II. British racing was isolated for the remainder of the war, although the 1944 Oaks winner Hycilla (by Hyperion) was not then eligible for the General Stud Book on account of her American dam Priscilla Carter. The end of the war exposed the British Thoroughbred once more to foreign competition, particularly from France, where the influence of Tourbillon and Djebel had become prominent. Before long the defences of 'purity of blood' were crumbling under the strain of persistent successes by half-breds. The crisis came in 1948, when two of the five Classic races were won by horses ineligible for the Stud Book. My Babu, the winner of the 2,000 Guineas, was by Djebel. Black Tarquin, the winner of the St. Leger, had as his fourth dam Frizette, by the unacceptable American sire Hamburg. Frizette was also the great-granddam of Tourbillon.

The defences of 'purity of blood' were down. The Advertisement to Volume XXXI of the General Stud Book, published in 1949, contained the following paragraphs:

'In July, 1948, the Publishers referred to the Stewards of the Jockey Club the question whether steps should now be taken to broaden the scope of the Book so as to allow for out-crossings with certain strains which at present were not admissible. The Stewards appointed a Committee to take evidence and report

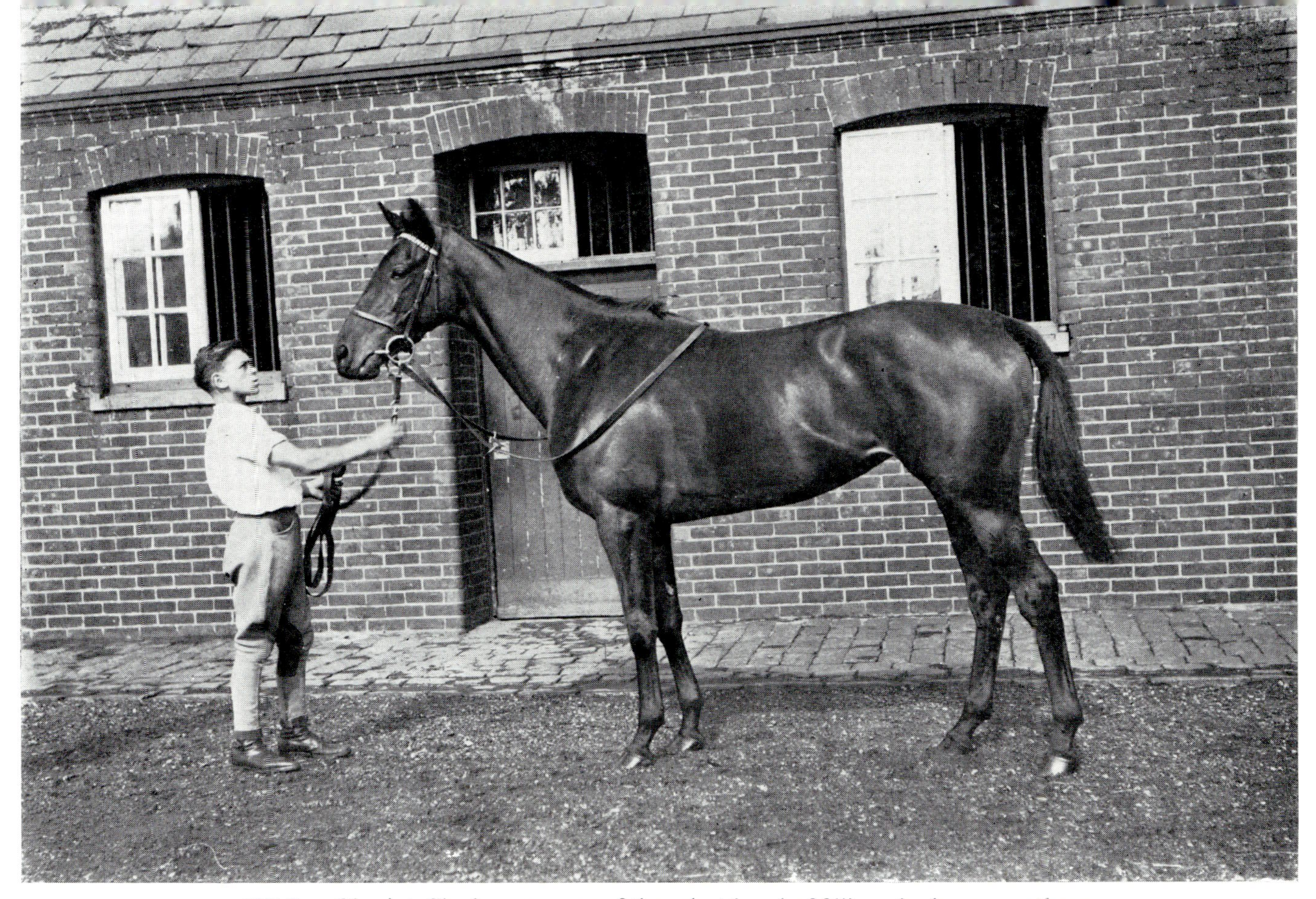

(25) Sun Chariot. She became one of the select band of fillies who have won three Classic races by winning the 1,000 Guineas, Oaks and St Leger in 1942.

*(26)* Sir Ivor. Foaled in 1965, he won the 2,000 Guineas and the Derby and was considered a perfect model of the thoroughbred

*(27)* Nijinsky. Foaled in 1967, he was the first winner of the British Triple Crown for thirty-five years

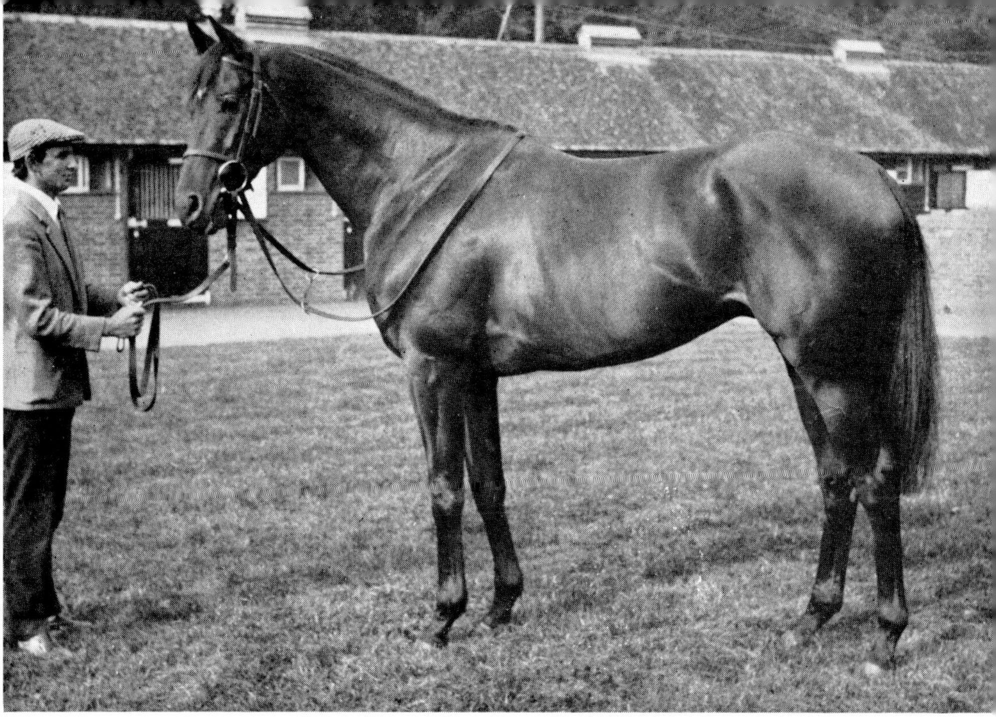

*(28)* Brigadier Gerard. Foaled in 1968, he won the 2,000 Guineas and sixteen of his other seventeen races. He was a horse of faultless conformation

*(29)* Ribot. The unbeaten winner of sixteen races, Ribot, bred in Italy in 1952, was one of the great racehorses and sires of the twentieth century

*(30)* Petite Etoile. A descendant of Mumatz Mahal, Petite Etoile combined brilliant speed with enough stamina to win the 1,000 Guineas and Oaks in 1959

upon the point, and the report given to the Jockey Club at their December meeting stated that some modification was advisable in the qualifications at present required.

'The Publishers, therefore, give notice that as from this date the conditions which have governed admission continuously since Volume XXII are rescinded.

'Any animal claiming admission from now onwards must be able to prove satisfactorily some eight or nine crosses of pure blood, to trace back for at least a century, and to show such performances of its immediate family on the Turf as to warrant the belief in the purity of its blood.'

This was a return to the conditions and the wording of 1901. By a stroke of the pen the stigma of illegitimacy had been removed from the vast majority of American thoroughbreds and their descendants in other countries, notably France.

Within fifteen years of the repeal of the Jersey Act the Derby had been won by four horses—Galcador, Never Say Die, Larkspur and Relko—who would have been ineligible for the General Stud Book if the Act had remained in force; and the St. Leger also had been won by four horses—Never Say Die again, Aurelius, Hethersett and Ragusa—who would have been debarred. The repeal had come just in time to save the face of the British Thoroughbred.

Twenty years after the repeal of the Jersey Act an attempt was made to rationalise the conditions of eligibility for the General Stud Book and to eliminate the scientific inaccuracy involved in the phrase 'purity of blood'. The revised conditions were applied to Volume XXXVI, published in 1969, and were as follows:

Any horse claiming admission to the General Stud Book should be able:

(1) To be traced at all points of its pedigree to strains already appearing in pedigrees in earlier volumes of the General Stud Book, these strains to be designated 'thoroughbred', or

(2) To prove satisfactorily eight 'thoroughbred crosses consecutively including the cross of which it is the progeny and to show such performances on the Turf in all sections of its pedigree as to warrant its assimilation with 'thoroughbreds'.

Thus at last the word 'thoroughbred' received an official definition some two hundred years after its first appearance in the terminology of the racehorse.

Volume XXXVI contained an important admission under

condition (2), that of Lavant. Lavant was a member of the famous 'half-bred' family of Verdict, the high class mare who won the Cambridgeshire in 1923 and the Coronation Cup the next year. Verdict bred the Oaks winner Quashed and Thankerton, who was third in the 2,000 Guineas and the Derby, and Verdict's great-granddaughter Lavant bred the two champion sprinters Lucasland and So Blessed. The closest mare of untraced origin in the direct female ancestry of Lavant was the mare by Perion, foaled about 1837, who was the seventh dam of Verdict. There is no doubt that Lavant qualified fully for assimilation with thoroughbreds.

Lavant blazed the trail. Members of other well-known 'half-bred' families, like those of Solerina and Hyacinthia Girl, were admitted to Volume XXXVII.

Non-thoroughbreds account for a large number of horses in training in Great Britain; indeed in October 1970 it was estimated that 10·2 per cent of all horses in training for jumping and flat racing in Great Britain were not thoroughbred. This created an anomalous situation, because Great Britain and Ireland were almost the only important racing countries in which horses of incomplete or unverified pedigree were permitted to race regularly against thoroughbreds, while regulations for the documentation and identification of thoroughbreds were being made ever more strict to meet the requirements of international racing and breeding. Accordingly the Jockey Club, on the recommendation of a committee headed by Mr. Edward Courage, introduced a register of non-thoroughbred mares which would be kept to the same stringent standards as the General Stud Book. This register came into operation, with more than 2,000 mares, on January 1st 1974, and no non-thoroughbred foaled after that date would be permitted to race unless registered. The register not only terminated the unsatisfactory double standard of documentation of racehorses, but provided the evidence required for future admission of non-thoroughbred strains to the General Stud Book when they fulfilled the conditions for entry.

### *Conclusion*

'Purity of Blood' had been a fine, resonant slogan, but the facts of life in the world of international racing after 1945 gave it a body blow. Perhaps it had never made a great deal of sense, granted the contradictions that were built into the Jersey Act and the acceptance of the dubious strains admitted to the General Stud Book before 1913. The results of the Tourbillon – Djebel era made the haughty exclusiveness of the period between the wars seem, in retrospect, rather pathetic.

A closed-shop stud-book policy is tenable where the animal concerned is not prized for food production, or capacity for useful work, or high standards of physical performance. Breeds of Pekinese, which come into none of these categories, need never fear extinction unless breeders pursue calamitously inept methods of selection or the dogs lose favour completely with dog-owners, and of these two dangers the latter seems infinitely remote. On the other hand, breeds of poultry, cattle and draught animals can survive only as long as their standards of production or work remain competitive with rival breeds, and, in the case of draught animals, with machines designed to do similar work.

In this respect the thoroughbred belongs with the poultry, cattle and draught animals, not with the Pekinese. No breed of racehorses, entrenched within a stud book based on 'purity of blood', has a chance of survival if its representatives are unable to hold their own in competition with other breeds, and insistence on 'purity of blood' regardless of racehorse performance is about as near as you can get, figuratively speaking, to flogging a dead horse. The General Stud Book, or any other stud book of a breed of racehorses, is doomed if it is regarded as an aspect of revealed and immutable truth. It must be treated realistically, be permitted to grow and develop, and to absorb its rivals when it cannot defeat them.

These reflections lead directly to two final questions: Was the Jersey Act inspired only by a combination of self-interest, stiff-necked insular pride and blind adherence to tradition? Was it pure folly? A positive answer to both those questions would disregard one of the principal purposes of a stud book, which is to ensure that animals included in it breed true, that is, are genetically similar. Genetic similarity is not identical with outward resemblance, and animals that appear to belong to the breed may carry harmful recessives; indeed animals that not only look thoroughbred but show superior qualities as racehorses may carry recessives that would endanger the future of the breed if they were admitted to the General Stud Book.

The other purpose of a Stud Book is that stated by James Weatherby in the original volume—to correct the evil of false and inaccurate pedigrees; and, conversely, to provide means of positive identification of horses and their breeding. Obviously certain standards must be maintained, lest these purposes be defeated. The case for a compromise on the lines of the existing conditions for admission to the General Stud Book is overwhelming.

# 10

# *International Competition and the Thoroughbred in the Second Half of the Twentieth Century*

THE PREFACE TO THE FIRST EDITION of Volume II of the General Stud Book, published in 1821, contained this passage:

'If any proof were wanting of the superiority of the English breed of horses over that of every other country, it might be found in the avidity with which they are sought by foreigners. The exportation of them to Russia, France, Germany etc., for the last five years, has been so considerable, as to render it an object of some importance in a commercial point of view.'

By the time Volume IV was published fifteen years later the list of countries to which stallions had been exported had swelled to include the United States, Baden, Mecklenburg, Holstein, Prussia, India, Sardinia, Van Diemen's Land,* Jamaica, Sweden and New South Wales. Racing was becoming a popular sport in many parts of the world, and most countries in which it took root began to found breeding industries to supply the raw material of the sport. Unless the British Isles enjoyed unique advantages of pasture and climate, or unless British breeders possessed unique expertise and refused to export stock of merit, some of those foreign countries were bound sooner or later to produce racehorses capable of challenging British supremacy.

### *Anglo-French Rivalry*

The first foreign horse to win an important English race was the German-bred Turnus, who made a successful raid on the Stewards Cup and the Chesterfield Cup at Goodwood in 1850. But Germany, partly on account of poor environment and partly

* Tasmania

on account of mistaken criteria of selection, failed to build on this early success, and it was left to England's nearest continental neighbour, France, to prove that a breeding industry on foreign soil was capable of mounting a sustained threat to the best products of Great Britain.

It would be too facile to lay stress on the irony that it was an Englishman, the eccentric Lord Henry Seymour, who played a decisive part in the development of racing and breeding in France. Nevertheless it is true that Seymour imported the English horse Royal Oak (after whom the French St. Leger was named) to his stud at Sablonville; that Royal Oak sired Poetess, who in 1841 became the last of Seymour's four French Derby winners: that Poetess was one of the greatest broodmares in the annals of the French Turf and was the dam of Monarque, who won the French 2,000 Guineas and the French Derby in his native country and came to England to win the Goodwood Cup in 1857: and that Monarque sired Gladiateur, who was acclaimed the 'Avenger of Waterloo' and destroyed the myth of the unchallengeable supremacy of the British Thoroughbred by winning the English Triple Crown in 1865.

Gladiateur's Triple Crown symbolised the French breeding industry's attainment of parity with the British. Indeed the French seemed not merely to have attained parity but to have surpassed the standards of the creators of the thoroughbred during the years which may be described as the 'Gladiateur era'. Gladiateur himself vied with St. Simon and Ormonde, both foaled years later, for the title of greatest horse of the century. His Classic triumphs had been preceded by the Oaks victory of Fille de l'Air the previous year, which was the first Classic victory of a French-bred horse in England; within the next fourteen years victories by French-bred horses in English Classic races were gained by Chamant in the 2,000 Guineas, Reine and Camelia in the 1,000 Guineas, Rayon d'Or in the St. Leger, and Reine, Enguerrande and Camelia in the Oaks, the two last-named running a dead-heat at Epsom in 1876. The complaint that the British Thoroughbred was degenerate was heard for the first time in the 1860s and 1870s. If Gladiateur had gone to stud in France and become as great a sire as he was racehorse, the era of French superiority might have been prolonged indefinitely. Instead he spent the whole of his short stud career except one season in England and died as a complete failure at the age of fourteen. Before long the pendulum swung England's way. In

1880 the English horse Robert The Devil won the Grand Prix de Paris, the principal French Classic race at that time, and the English horses Bruce, Paradox and Minting followed up by winning the Grand Prix three times more in the next six years. The improvement of English fortunes reached its climax with St. Simon and Ormonde, and the incomparable St. Simon ensured that the great English races would not be vulnerable to foreign invasion for an equine generation. Racing was suspended in France during the First World War while the sport continued, though severely curtailed, in England. The consequence was that the French were at a grave disadvantage immediately after the war and the British Galloper Light, Comrade and Lemonora exploited the favourable situation to win the Grand Prix three years in succession before the recovery of French breeding was complete.

The reasons for the temporary superiority of the products of one country or the other are readily discernible in certain periods. The skill and initiative of Gladiateur's owner-breeder Count Frédéric de Lagrange were largely responsible for the first flush of French victories. St. Simon restored British supremacy. The suspension of the racecourse test and the disorganisation of the French breeding industry clearly accounted for the French defeats between 1919 and 1921. French superiority after the Second World War depended to no inconsiderable extent on the brilliant success of the breeding formula evolved by the leading French owner-breeder Marcel Boussac, using the prepotent sires Tourbillon, Pharis II and Asterus in various combinations; also on the adverse effects of Jersey Act exclusiveness which denied British breeders access to a rich seam of racing merit.

During the twenty-year period from 1946, when full scale international competition was resumed after the Second World War, to 1965 twenty-two of the hundred Classic races run in Great Britain were won by French horses. This total comprised three victories for French horses in the 2,000 Guineas (My Babu, Thunderhead II and Niksar), three victories in the 1,000 Guineas (Imprudence, Camaree and Bella Paola), seven victories in the Derby (Pearl Diver, My Love, Galcador, Phil Drake, Lavandin, Relko and Sea Bird II), six victories in the Oaks (Imprudence, Asmena, Sun Cap, Sicarelle, Bella Paola and Monade) and three victories in the St. Leger (Scratch II, Talma II and Cambremer). These victories in the Classic races were supplemented by a similar success ratio for French horses in other British tests for top-class horses like the Ascot Gold Cup, the Eclipse Stakes, the King George VI and Queen Elizabeth Stakes and the Champion Stakes.

After 1965 the tide of French success in Great Britain turned and receded abruptly. Indeed only two French horses won English Classic races between 1966 and 1973, and they were Valoris (1966) and La Lagune (1968) who both won the Oaks. During the same period horses bred in Great Britain and Ireland—the breeding industries of the two countries are closely integrated and share the same General Stud Book—invaded the French Classic races with such effect that three won the French 2,000 Guineas (Zeddaan, Caro and Kalamoun), one won the Frnech Oaks (Sweet Mimosa), two won the French Derby (Sassafras and Hard To Beat), two won the Grand Prix (Roll of Honour and Pleben) and three won the French St. Leger (Samos III, Sassafras and Pleben).

This was a dramatic reversal of fortune. The reasons for it were certainly complex, but one of them, and not the least telling, is so obvious that it may be overlooked. No horse can win a race unless he is granted the opportunity to do so, and successive increases in French prize money, in an absolute sense and in relation to prize money in Great Britain, greatly diminished the incentive for French horses to run in the British Classic and other important races during the 1960s. Conversely, British horses had an ever-growing incentive to run in France.

Nevertheless the shift in the balance of prize money cannot be held wholly responsible for the change. Classic races confer prestige and enhanced capital value on the winners independently of prize money, and horses will cross seas and frontiers to win them, even if only to escape stronger opposition at home, whenever the opportunities are ripe. If French-bred horses had retained a general superiority, British owners would have flocked to buy them as foals or yearlings in an attempt to win important races at home. Undoubtedly there was a change in the relative prowess of British thoroughbreds on the one hand and French thoroughbreds on the other hand after 1965.

Is this change susceptible of analysis and explanation, or must it be interpreted purely in terms of an inexplicable swing of the pendulum? One thing is certain: the Tourbillon – Pharis II – Asterus formula lost its efficacy in the same way that a seam of ore may be exhausted or a region lose its fertility through over-cultivation or the caprices of climate. The Boussac breeding empire collapsed nearly ten years before the decline of French breeding as a whole became apparent. The results of French Classic races told the story. Marcel Boussac won the French Derby with Philius and the French Oaks with Apollonia in 1956, and his reputation as Europe's master breeder seemed as secure as ever. But events were to prove that the Boussac studs were on the brink of severe retrogression; they did not produce another Classic winner until Crepellana won the French Oaks thirteen years later, and this filly was the result of an outcross and was by the British stallion Crepello.

Powerful as the Boussac studs were at the peak of their fame, spearhead as they did the French assault on the great British races, they never represented the whole front-line strength of French breeding. They provided twelve winners and placed horses in British Classic races in the twenty years after the end of the Second World War—a marvellous achievement but not one that denoted supremacy. The decline in the rate of French success after 1965 can be explained only in terms of a general French recession, or of a regeneration of British breeding, or a combination of both.

Several different factors probably contributed to the decline of French thoroughbred standards. They include too much emphasis on stamina to the neglect of the vital attribute of speed: a rate of exportation of the country's best stallions and mares that depressed the overall quality of French bloodstock: a failure to requite these exports with imported bloodstock of comparable quality: excessive watering of tracks to the detriment of the racecourse test for speed, action and soundness: and lack of insistence on soundness and constitution in the methods of selection practised in some key studs.

### *The Thoroughbred in North America*

There have been continuous racing contacts, interrupted only by two world wars, between Great Britain and France for a century and a quarter. The propinquity of the two countries, and the early foundation of a racehorse breeding industry in France, made Franco-British rivalry the most convenient field for the study of international bloodstock competition up to the 1960s. However this is not the only field, nor necessarily the most fruitful field, for the study of this kind of competition. The rivalry between the racehorse breeding industries of North

America and Europe may hold lessons of at least equal significance.

There was horse racing in North America from the early days of English colonisation, and men like Fenwick of South Carolina, Carter of Virginia, Tasker of Maryland and Delancey of New York were founding studs on American soil during the eighteenth century when the evolution of the British thoroughbred was still in its infancy. The importation of the first Derby winner Diomed, a failure at stud in England, gave a vital new impetus to the American breed, which was consolidated by the influence of other English stallions like Leamington, Glencoe, Australian and Eclipse (not to be confused with the earlier and even more celebrated Eclipse by Marske) imported during the middle years of the nineteenth century.

The first American-bred horses began to make their presence felt on English courses during the 1850s, when Prioress won the Cesarewitch and Starke won the Goodwood Stakes and the Goodwood Cup. In 1881 Iroquois proved the American breeding industry capable of producing horses of the highest class by winning the Derby and the St. Leger, and Foxhall, by repute an even better horse but regrettably omitted from the entries for the English Classic races, won the Grand Prix de Paris and landed the coveted autumn double of the Cesarewitch and the Cambridgeshire. Foxhall won the Ascot Gold Cup the following year. Other American-bred horses Sibola, Cap and Bells II, Norman III, Sweeper II and Tracery won English Classic races before the First World War, and the Derby winners Orby and Durbar II had American-bred dams.

There was relatively little transatlantic competition between the two world wars, but it was resumed at a steadily accelerating pace after 1945. The American-bred Black Tarquin won the St. Leger in 1948, and six years later Never Say Die became the first American-bred Derby winner since Iroquois, whose achievement of adding victory in the St. Leger he also repeated. During the 1960s, which opened with the victories of the American-bred filly Never Too Late II in the 1,000 Guineas and the Oaks, the evidence of American thoroughbred superiority began to accumulate steadily, both in Great Britain and continental Europe. The results of the seven greatest races in Great Britain (the 2,000 Guineas, the 1,000 Guineas, the Derby, the Oaks, the St. Leger, the King George VI and Queen Elizabeth Stakes and the Champion Stakes) and the seven greatest races in France (the 2,000 Guineas, the 1,000 Guineas, the Derby, the Oaks, the St. Leger, the Grand Prix de Paris and the Prix de l'Arc de Triomphe) tell the story. During the period 1968–73 13 of the 42 (31 per cent) British races were won by American-bred horses, while 9 of the 42 (21 per cent) of the French races were won by American-bred horses. Four winners of the Derby (Sir Ivor, Nijinsky, Mill Reef and Roberto) were bred in North America, and two winners (Mill Reef and San San) of the Prix de l'Arc de Triomphe, the richest race in the world, were bred in the United States.

This American supremacy was even more evident in the tests of speed and precocity than in longer races for older horses. For example in 1973 all the five most important French two-year-old races were won by American-bred horses, Lianga winning the Prix Robert Papin, Nonoalco the Prix Morny and the Prix de la Salamandre, Hippodamia the Criterium des Pouliches and Mississippian the Grand Criterium; and four of the five most important English two-year-old races were won by American-bred horses, Gentle Thoughts winning the Flying Childers Stakes and the Cheveley Park Stakes, Cellini the Dewhurst Stakes and Apalachee the Observer Gold Cup. The only race to escape the net of American breeding was the Middle Park Stakes, whose winner Habat was bred in England, though he was by the American-bred stallion Habitat.

This unprecedented incidence of successes for American-bred horses in Europe has been rendered possible by the fact that they have sent over to be trained in Great Britain, Ireland and France in rapidly increasing numbers. But the scale of American success in the late 1960s and early 1970s cannot be explained solely in terms of opportunity, because the opportunities have arisen as a result of the growing awareness of American owners, and of European owners who have begun to invest heavily in American bloodstock, of the superior ability of many American-bred horses.

The history of the Washington D.C. International, an invitation race founded at Laurel, Maryland, in 1952, demonstrates not only the excellence of American horses but their relative improvement during the last quarter of a century. The results of the first 22 Internationals (1952–73) show fairly even scoring, with 12 winners bred in the United States and 10 bred elsewhere. But the results of the first seven and the last seven present a marked contrast; only one American-bred horse,

Fisherman, was successful in the initial period, whereas only one foreigner, the Irish-bred Karabas, was successful in the latest period while the American-bred horses Fort Marcy (twice), Sir Ivor, Run The Gantlet, Droll Role and Dahlia won the other six races.

*Some Reasons for American Progress*

The American breeding industry was able to produce horses like Iroquois and Foxhall capable of winning European Classic races as early as the final quarter of the nineteenth century. American bloodstock imported into Great Britain and Ireland about the turn of the century, for example Americus, Rhoda B and Sibola, had a profound influence on the later evolution of the British thoroughbred; and Frizette, imported into France from the United States at the end of the first decade of the twentieth century, had an equally profound influence on the later evolution of the French thoroughbred, particularly through her great grandson Tourbillon, who was one of the main factors in the French ascendancy of the 1940s and 1950s.

Nevertheless the influence of American breeding penetrated far more deeply, and the capacity of the American breeding industry consistently to produce top class horses by European standards became infinitely greater, during the quarter of a century after 1950. The question how this happened is crucial to any understanding of the mainsprings of thoroughbred progress.

The problem is complex and the answer cannot be simple. All that is possible is to suggest some of the contributory causes. The first is the sheer size of the North American breeding industry. In 1946 the number of foals born in North America (the United States and Canada) was 6,579; in 1970 the number had risen to 22,747, an increase of 250 per cent. During the same period annual foal production in Great Britain and Ireland rose from 4,106 to 6,361, an increase of only 55 per cent. This American expansion, not merely in absolute terms but in relation to the homelands of the thoroughbred, provided the broadest possible base for progress.

But quantity alone cannot guarantee improvement. Indeed if expansion means a lowering of standards of selection and the use of animals for breeding that ought to be culled, it may cause deterioration. The essential concomitant of expansion in American breeding has been heavy investment in the best bloodstock wherever it is available. In the period between the two world wars American breeders imported stallions like the brothers Sir Gallahad III and Bull Dog with most beneficial results. The rate of importation of top class stallions was maintained during the Second World War, when the depressed state of the market in Great Britain and Ireland facilitated the purchase of the Derby winners Blenheim, Mahmoud and Bahram at bargain prices. After the war the immense wealth of the United States enabled Americans to buy more aggressively than ever before, and stallions like Nasrullah, Royal Charger and Ambiorix, and mares like La Mirambule (dam of Tambourine II and Nasram II) and Gloria Nicky (dam of Never Too Late II), and countless other high class animals of both sexes were imported from countries as widespread as Great Britain, Ireland, France, Germany, Italy, Argentina and Australia. Inevitably some of these imported animals, however illustrious their lineage or brilliant their racing performances, were failures at stud, but the cumulative effect was to raise American thoroughbred standards in dramatic fashion.

The leader in this intensive investment programme was A. B. (Bull) Hancock, the owner of the most famous of all American studs, Claiborne Farm at Paris, Kentucky. Hancock's father Arthur B. Hancock had imported Sir Gallahad III, who was leading sire of winners four times and leading broodmare sire twelve times. Bull Hancock made a close study of the most successful bloodlines all over the world, and set out to introduce potent strains that were absent or scarce in the United States. He found that the male line of Nearco was weakly represented, so sought out and purchased his son Nasrullah, who became the leading sire of winners five times. A stud companion of Nasrullah at Claiborne was another stallion of European origin but of completely different pedigree and aptitude, Princequillo, who was leading sire of winners twice. Together they formed one of the strongest 'nicks' in the history of the thoroughbred. Nasrullah's son Bold Ruler, who was champion sire eight times, and Princequillo's son Round Table, who was champion sire for the first time in 1972, both stood at Claiborne and combined to project the Nasrullah – Princequillo influence into later generations.

Hancock's restless search for stallions representing successful bloodlines in various parts of the world led him to buy Ambiorix, Herbager and Le Fabuleux from France; Tatan, Pronto and Forli from Argentina; and Sky High II and Pago Pago from Australia—among many others.

Superimposed on this investment programme was a rigorous system of selection based on the racecourse test. American horses, irrespective of their class, are expected to race hard and often. Bold Ruler and Princequillo each ran 33 times, and won 23 and 12 races respectively. Round Table ran exactly twice as many times as his sire, and won 43 races. Successful British thoroughbreds tend to have much easier racing careers. The five English-bred winners of the Derby between 1964 and 1973 (Santa Claus, Charlottown, Royal Palace, Blakeney and Morston) took part in an average of only 8·4 races each.

But the number of times top-class horses run is not the only meaningful aspect of the racecourse test. Aspects of perhaps even greater significance are the distances of the races in which they run and their ages when they run in those races. The system of racing, which means the whole programme of races intended for the best horses, in any country will determine the type of horse produced in that country. For example, a country in which all the most valuable and prestige-bearing races were sprints for two-year-olds would tend to produce only precociously speedy horses of severely limited stamina, because those races would provide the criterion of excellence, and consequently the basis of selection for breeding.

Comparison of the racing systems of France and Great Britain is made easy by the existence of the internationally agreed 'Pattern of Racing'* which incorporates the races designed for the thorough testing of the best horses. The British 'Pattern' in 1973 contained 28 races over 5 and 6 furlongs, and of these 15 were confined to two-year-olds; whereas the French 'Pattern' contained only 11 races over those distances, and no more than 4 of them were confined to two-year-olds. On the other hand the French 'Pattern' contained 48 races for three-year-olds and upwards over distances from 10 to 15½ furlongs, compared with 38 such races in Great Britain.

The inference must be that the British Pattern gives a much stronger incentive for the breeding of sprinters than the French; and indeed British-bred horses achieved almost a monopoly of success in the French Pattern sprint races in the late 1960s and early 1970s. The Prix de l'Abbaye, run over 5 furlongs at Longchamp on the same day as the Prix de l'Arc de Triomphe, is the

* See Glossary—Pattern Races.

most valuable sprint race in France; and it was won by a horse bred in Great Britain or Ireland—the winners were Silver Shark, Farhana, Pentathlon, Be Friendly, Tower Walk, Balidar, Sweet Revenge, Deep Diver and Sandford Lad—each year from 1965 to 1973.

Conversely the French Pattern encourages the breeding of horses likely to shine over middle distances at three years of age and above, and it was in precisely this category that French horses achieved their greatest degree of superiority during the twenty years after the Second World War.

Comparison with the American system cannot be so direct because there is no official Pattern in the United States*. There has to be recourse to more general criteria. The opinion, widely held in European racing circles, that the American system gives powerful, and indeed exclusive, encouragement to precocious speed is not accurate, and can hardly be reconciled with the fact that many American horses have been successful in European Classic and other important middle distance races. That there is strong emphasis on speed is undeniable. Races over distances up to 6 furlongs accounted for 56·57 per cent, and races over distances up to a mile accounted for 81 per cent, whereas races over 1½ miles and beyond accounted for only 0·39 per cent of the 53,227 races run in North America in 1972. But it is the programme of races for the best horses (the 'Pattern' in the British sense of the word) that has the decisive impact on selection, and quality gets a much fairer share of prize money in longer races. For example the average first prize for the 107 races over 1½ miles was 21,406 dollars, more than six times as much as the average first prize for the 18,693 races over 6 furlongs in North America in 1972.

The series of big American two-year-old races begins with the Sapling Stakes and the Arlington-Washington Futurity (both 6 furlongs) and the Hopeful Stakes (6½ furlongs) during August, continues with the Belmont Futurity (6½ furlongs) in September, the Champagne Stakes (1 mile) and the Laurel Futurity (8½ furlongs) in October, and concludes with the Garden State Stakes (8½ furlongs) in November. The American Triple Crown races, comprising the Kentucky Derby (1¼ miles), the Preakness Stakes (9½ furlongs) and the Belmont Stakes (1½ miles)

*A step towards an official American Pattern was taken with the publication of the pamphlet 'North American Graded Stakes Races' by the Thoroughbred Owners and Breeders Association in 1974.

are run in a period of six weeks from early May to mid-June.

Secretariat, who was to win the Triple Crown the next year, won the Hopeful Stakes, the Belmont Futurity, the Laurel Futurity and the Garden State Stakes in the course of his two-year-old campaign. And it is significant that the first prize of 179,199 dollars credited to him in the Garden State Stakes was much larger than the sum of 140,300 dollars credited to his stable companion Riva Ridge for winning the premier American Classic race, the Kentucky Derby, the same year, 1972.

It is true that the American system gives a powerful incentive to the development of precocity, but not precocity in the sense of ability to scamper 5 furlongs or less in the first half of the two-year-old season. The ideal of the American breeder is a horse capable of outrunning the best of his contemporaries over distances ranging from 6 furlongs in August as a two-year-old to 1½ miles in mid-June as a three-year-old, with the richest reward assigned to excellence over a mile or a little more in the autumn of the former season.

The American system does indeed place stronger emphasis on a combination of speed and precocity than either the British Classic series, which enshrines the ideal of a horse superior to his contemporaries over a mile in April/May, 1½ miles in early June and 1¾ miles in September of his three-year-old season, or the French system with its monolithic insistence on middle distance racing. But the British system is flawed by its ambivalence, which splits the resources of the breeding industry between two irreconcilable types, the early maturing sprinter bred to race over 5 and 6 furlongs as a two-year-old and the Classic type maturing a year later. The emphasis in the American system is perfectly compatible with the production of horses able to complete with the best European horses not only at two but at three years of age, and the promotion of an industry able to produce such horses regularly and in substantial numbers has been the triumph of American breeders in the second half of the twentieth century.

### *International Breeding and Hybrid Vigour*

The Sailor Prince, the winner of the Cambridgeshire under a low weight in 1886, was exported to the United States as a stallion. There he was mated with Saluda, whose family had been in North America for generations and whose granddam Maiden was by the greatest of nineteenth-century American stallions, Lexington. The produce of this mating was Sibola, who was sent to England to win the 1,000 Guineas and finish second in the Oaks in 1899.

Catnip, a daughter of Sibola by Spearmint (by the New Zealand-bred Carbine) was bought by the Italian breeding genius Federico Tesio and taken to Italy. There she was mated with Havresac II, who was inbred to St. Simon in the second and third generations. The produce was Nogara, a winner of the Italian 1,000 and 2,000 Guineas. Nogara was mated with Pharos, who was inbred to St. Simon in the third and fourth generations, and the produce was Nearco, unbeaten in 14 races, one of the most brilliant racehorses and most potent stallions of the twentieth century.

Nearco spent his stud career at the Beech House Stud at Newmarket. One of the best of his sons was Nasrullah, whose granddam Mumtaz Mahal was so fast that she was nicknamed the 'Flying Filly'. She was by the Irish-bred prodigy The Tetrarch (by the French-bred Roi Herode), but her graddam Americus Girl was by the American-bred sprinter Americus.

Nasrullah, talented but wilful on the racecourse, was leading sire of winners once in Great Britain and Ireland, but his results did not convey the impression that he would ever attain the status of a great stallion in those countries. Purchased by Hancock and transported to Kentucky, he became as potent a stallion as his own sire. The most famous and influential of his sons, Bold Ruler, was out of Miss Disco, who had no imported ancestor later than the third generation of her pedigree and sprang from the male line of Australian, which had flourished in North America since the middle of the nineteenth century.

The obvious conclusions that may be drawn from the process of heredity linking Saluda and Bold Ruler are that thoroughbred strains may be transferred from country to country and continent to continent without loss of quality, and that the highest racing class may be obtained by mingling strains derived from many different sources. Analysis of the successive stages of the process suggest that further, perhaps even more significant, lessons may be learnt. The Sailor Prince, mated with an American mare, produced in Sibola a filly of a class superior to his own. Catnip, mated with a stallion inbred to a prepotent sire entirely absent from her own pedigree, produced in Nogara a racehorse and broodmare of brilliant attainments. Nearco, mated with a mare whose pedigree included exceptionally diverse elements, sired in Nasrullah a wonderfully gifted horse; but Nasrullah fulfilled his potential excellence as a stallion only when he was removed from his native country and mated with a mare drawn from strains unrelated to his own.

This analysis leads to an important conclusion. It is that hybrid vigour in the thoroughbred may be engendered in two different sets of circumstances. Firstly, this phenomenon may arise simply from the mating of unrelated individuals, particularly if one is inbred to an prepotent ancestor, or if both are so inbred. Secondly, hybrid vigour may be induced, or at least enhanced, when strains developed in contrasting climatic or geographical environments are united.

This second kind of hybrid vigour acts as a bonus to the American policy of acquiring the best bloodstock wherever available. American breeders have profited not only from accumulating the best from all over the world, but also from this unbudgeted factor which springs from crossing strains developed in different environmental conditions.

The extremely wide geographical spread of the North American breeding industry itself provides the contrasts needed to provide this additional source of thoroughbred progress. Regions as diverse as the semi-desert and the mediterranean areas of California in the west, temperate Kentucky, Virginia and Maryland in the east, semi-tropical Florida in the south and Canada, with its long cold winters and humid summers, in the north have all been able to produce racehorses of the highest class; and their integration in a single breeding industry—the thoroughbred operations of Canada and the United States are indivisible—gives an important advantage to the thoroughbred in North America.

### *The Future of the Thoroughbred*

An annual rate of foal production unmatched elsewhere: an unprecedented concentration of quality due to heavy and sustained investment in the best bloodstock wherever available: rigorous application of the racecourse test: a favourable system for top class horses: and the bonus of hybrid vigour—all these factors that have helped to promote the superiority of the American thoroughbred are likely to operate for an indefinite time to come. A new development, drawing inspiration from this American superiority, is the growth of large-scale international racing and breeding enterprises which is gathering momentum in the early 1970s. The American Nelson Bunker Hunt, the Japanese Zenya Yoshida, the French art dealer Daniel Wildenstein and the Swiss-domiciled Countess Margit Batthyany were among the leaders in the initial stages of this development.

The increased competition generated by these enterprises has intensified the inflationary pressure on values of the best and most fashionable bloodstock. The result is that breeders in Great Britain, where prize money is low by the standards of the principal racing countries and tax burdens are onerous, have been placed at a grave disadvantage against breeders in the nearest European country France, where high prize money is complemented by generous breeders' premiums.

French breeders have been able to replenish their resources of high class bloodstock freely. In 1954 the French pool of stallions included only 32 that had won Group I (the championship grade in the International Pattern) Pattern races; twenty years later the number of such horses had risen to 52. In Great Britain the number of Group I winners in the stallion pool rose only from 49 to 54 in the same period. Thus the quality of the French breeding industry, which measured by the number of mares is only about two-thirds the size of the British—Ireland is excluded from this calculation—has made tremendous relative improvement.

On the other hand this quality improvement had not been reflected in racing reults by 1973, when the prize money won by foreign-bred horses in France reached the record level of 35·42 per cent of the total distributed. This failure of French-bred horses to hold their own at home led to the introduction of restrictions on foreign-bred and trained horses running in France during 1974, though happily the Pattern Races were immune from these protective measures.

But the days when international competition could be seen in the crude terms of 'French and English' have gone for ever. Other countries besides the United States, notably Japan, have invested heavily in the choicest bloodstock. The horse racing business is world-wide. The North American breeding industry may have achieved dominance, but the demarcation lines of the various national breeding industries have been blurred more and more by international breeding operations. These sweeping changes offer new and exciting hopes of progress in thoroughbred performance.

# *Glossary*

Some of the following words and phrases have appeared in the text, though without adequate explanation, and others have not. All are in common use in the phraseology of the thoroughbred.

*Brothers and Sisters and other relationships*

Brothers and Sisters, otherwise called Full Brothers and Sisters or Own Brothers and Sisters: These are animals having the same sire and dam. Thus Dante and Sayajirao were brothers because they were both by Nearco out of Rosy Legend; and Pharos (male), Fairway (male) and Fair Isle (female) were brothers and sister because they were all by Phalaris out of Scapa Flow.

Half Brothers and Sisters: Animals out of the same dam but by a different sire. Thus Sheshoon, by Precipitation out of Noorani, was a half-brother of Charlottesville, by Prince Chevalier out of Noorani.

Threequarters Brothers and Sisters, otherwise called Three Parts Brothers and Sisters: Animals out of the same dam and having as their sires animals who were half-brothers, in the sense described above, or were by the same sire but out of a different dam. Thus Neasham Belle, by Nearco (by Pharos) out of Phase, was a three-quarters-sister of Setting Star, by Signal Light (by Pharos) out of Phase; likewise Mortification, by Precipitation (by Hurry On out of Double Life) out of Miss Pecksniff, was a three-parts brother of Persepolis, by Persian Gulf (by Bahram out of Double Life) out of Miss Pecksniff.

*Chance-bred*

This term refers to an animal that shows high-class form but has one or both parents of poor pedigree and performance. The assumption is that a chance-bred animal is unlikely to have inherited many of the genes making for good performance from both parents, and so is unlikely to hold many of those genes in double strength. The inference is that chance-bred animals are less likely to be prepotent, and therefore valuable for breeding, than animals both of whose parents had good pedigree and performance.

*Classic Races*

In England the Classic races are the 1,000 Guineas and the Oaks for three-year-old fillies, and the 2,000 Guineas, Derby and St. Leger for three-year-old entire colts and fillies. The corresponding races are the Classic races in Ireland.

The Classic races in France, for the purposes of international comparison, are the Poule d'Essai des Pouliches (1,000 Guineas) and the Prix de Diane (Oaks) for three-year-old fillies, the Poule d'Essai des Poulains (2,000 Guineas) for three-year-old entire colts, and the Prix du Jockey Club (Derby) and the Prix Royal Oak (St. Leger) for three-year-old entire colts and fillies, with the Grand Prix de Paris for three-year-old entire colts and fillies in addition.

*Dual Paternity*

This term applies to the pedigree of an animal whose dam has been covered by two different stallions during the covering season in which he was conceived. In the written pedigree the animal is described as being by X or Y, X being the first and Y the second stallion by whom the mare was covered. It is assumed that the mare failed to conceive to the mating with X and that Y is the true sire of the produce. Supreme Court, the winner of the King George VI and Queen Elizabeth Festival of Britain Stakes in 1951, was described as 'by Persian Gulf or Precipitation', though there is no real doubt that Precipitation was the true sire.

It is expected that most problems of dual paternity will be solved as a result of research into blood typing of horses financed by Weatherbys at the Equine Research Station of the Animal Health Trust at Balaton Lodge, Newmarket. The Research

Station's report for the year ending 30th June 1973 stated that approximately 450 individual thoroughbreds had been typed and added:

'We have a very comprehensive set-up for thoroughbreds, using routinely seven electrophoresis systems plus fifteen red-cell grouping reagents comprising seven systems. This compares very favourably with other laboratories in the world, and probably leads in the field of thoroughbred blood typing. We have been able to calculate the frequencies of the various blood types and from these figures to arrive at an estimate of the probability of making an exclusion in cases of double covering etc: this is in excess of 75 per cent.'

*Fertility*

Low fertility is one of the most serious problems of the bloodstock industry. 'The Return of Mares' published by *The Statistical Record* showed that of 16,771 mares accounted for in 1973 only 8,378, or 50 per cent, produced live foals. However this percentage does not give an accurate picture of the state of fertility in the thoroughbred, because the total of mares accounted for included 1,920 mares that were not covered for one reason or another. Of the 14,581 mares actually covered 9,528, or 64·15 per cent, conceived. Even this figure must be regarded as highly unsatisfactory and wasteful.

Any significant improvement in fertility would being great economies to the industry. It would also give the options of increasing production from the same number of stallions and mares or of improving quality by culling the poorer specimens of the breed of both sexes. The potential benefits of improved fertility were so obvious that the joint committee of the Thoroughbred Breeders Association and the Hunters Improvement and National Light Horse Breeding Society initiated a research project in equine reproduction. This research, financed to a large extent by the Horserace Betting Levy Board and launched on 1st January 1969, was undertaken by Dr. W. R. Allen, using facilities made available by the Animal Research Station of the Agricultural Research Council at Cambridge.

One aspect of the problem to which Dr. Allen directed a detailed investigation is early foetal death. It has been estimated that in as many as 8–10 per cent of all thoroughbred mares covered during a season the foetus dies within the first 80–100 days of gestation and the products of pregnancy are either resorbed by the mare or aborted. One of Dr. Allen's most promising discoveries was the possibility of controlling the oestrus cycle of mares through the compound Prostaglandin (PG) and thereby overcoming many of the difficulties resulting from early foetal death and various irregularities of the oestrus cycle.

Other causes of low fertility are infection of the uterus, infective abortion, difficulties with foaling and problems of rearing foals. The whole subject of equine reproduction has proved exceedingly complex and it has become clear that no single solution to the problem of infertility in the mare can ever be found.

*Pattern Races*

The Pattern Races compose a system of races designed for the thorough testing of the best horses of all ages. They are the races on which selection of horses for breeding is based, and therefore have a vital impact on thoroughbred evolution.

The need for an officially promoted Pattern was specified in the report of the Duke of Norfolk's Pattern of Racing Committee in 1965. Two years later the British Pattern was introduced after detailed recommendations had been made by Lord Porchester's Race Planning Committee, which was then appointed a permanent standing committee to watch over the integrity of the Pattern.

The British initiative led to the creation of an International Pattern by agreement of the Turf authorities of Great Britain, France and Ireland. Pattern races were divided into three groups. Group I comprised the Classic and other championship races in which horses meet at level weights or strict weight-for-age terms. Group II comprised the principal supporting races whose conditions may incorporate some penalties and allowances. Group III comprised the other important races, including Classic trials, which are of mainly domestic interest in each of the participating countries.

In 1973 there were 100 Pattern races, including 19 Group I races, in Great Britain: 99 Pattern races, including 22 Group I races, in France: and 22 Pattern races, including 5 Group I races (the Classic races) in Ireland.

The International Pattern provides a criterion of the class of racehorses in different countries and is the most valuable guide to selection available to breeders.

*Syndication of Stallions*

The syndication of stallions has become common practice since the Second World War and become so widespread that it has, generally speaking, superseded the former system of standing stallions at a fixed covering fee as far as top-class stallions are concerned. The syndication of a stallion involves his capitalisation in forty shares—forty being the number of mares normally covered by a stallion in one season—and the offer of some or all of those shares for sale. Ownership of a share confers the right to send a mare to the stallion each season, or to dispose of the nomination in some other way. Management of the stallion is vested in a syndicate committee which normally has the right to sell one or more extra nominations in each covering season in order to defray the expense of keeping the stallion.

From the point of view of the original owner of the stallion, syndication has two advantages. It enables him to net a large capital sum—as international bloodstock values have inflated, the initial value of shares in top-class horses has risen to the £25,000 to £50,000 range—while retaining several shares which he may use for his own mares, swop for nominations to other syndicated stallions, or sell in the open market for income.

Syndication has operated to the advantage of owners of top-class horses going to stud. It is less certain that it has operated to the advantage of the British Thoroughbred. Essentially, syndication is a restrictive practice, and limits the chance of the stallion covering mares likely to produce good offspring as a result. A breeder owning a share may be tempted to use it even if he has no suitable mare; alternatively he may regard his share merely as a speculative counter and not as a means of ensuring that he has a nomination to a top-class stallion for the primary purpose of getting good horses. Large sums have been made by successful dealing in stallion shares. Some serious breeders prefer small partnerships of like-minded breeders to own stallions, in the belief that these provide better control and better opportunities for stallions to cover carefully chosen mares.

*Twins*

Twins usually result from the fertilisation of two individual eggs which have developed from separate follicles in the ovary. Identical twins, which result from the splitting of the egg into two parts soon after it is fertilised, are rare.

The uterus of the mare cannot easily accommodate twins or provide a proper environment for their development. Many twins are aborted or born dead. Modern veterinary techniques can often disclose that a mare is likely to shed two eggs during a given oestrus period, so she can be withheld from covering during that period. In 1972 66, or 0·8 per cent of the 7,919 mares returned as having foals to the General Stud Book produced live twins; and in 1973 105, or 1·2 per cent, of the 8,378 mares returned as having foals produced live twins. Owing to poor pre-natal environment, the twins that do survive tend to be weakly and undersized. Few twins do well on the racecourse and only one, the 1823 2,000 Guineas winner Nicolo, has won a Classic race. Nicolo's twin sister died at birth.

Nicolo went to stud and sired a number of winners, as did Trapeze, a twin and a younger brother of that horse of steel Tristan, who was second to Foxhall in the Grand Prix de Paris, and won the Ascot Gold Cup once and the Hardwicke Stakes three times. Many female twins have done well as broodmares. The most famous pair of twins in thoroughbred history were Lady Bawn and Lady Black, foaled in 1902. Both went to stud, and they produced five winners each. All Lady Bawn's winners were by Tredennis, and three of Lady Black's winners were by Tredennis. The best of the produce of either of these mares was Lady Bawn's son Bachelor's Double, winner of the Irish Derby, the City and Suburban, Royal Hunt Cup, and Kempton Park Great Jubilee Handicap. Bachelor's Double was a successful sire, getting the Grand Prix winner Comrade and the Oaks winner Love In Idleness, and was an outstanding sire of brood mares. His daughters bred the Derby winner Call Boy, the St. Leger winner Scottish Union, the Oaks winner Lovely Rosa, the 1,000 Guineas winner Pillion, the Grand Prix winner Fiterari and the Ascot Gold Cup winner Precipitation.

A more recent twin who has achieved fame as a brood mare is Dalila, the dam of Scissors, who finished first in the Timeform Gold Cup of 1963 but was disqualified in favour of Pushful.

Twins tend to suffer from an unfavourable environment during their pre-natal life. On the other hand a twin has as good a chance of inheriting a satisfactory genotype as a single offspring of the same mating. This is the reason why a twin, though undersized, may do well at stud.

# *Pedigrees*

Flying Childers (1715) and Bartlett's Childers (1716) (General Stud Book version) or Bleeding Childers

| | | | | |
|---|---|---|---|---|
| Darley Arabian | | | | |
| Betty Leedes | Old Careless | Spanker | D'Arcy Yellow Turk | |
| | | | Old Morocco Mare | |
| | | Barb Mare | | |
| | Cream Cheeks | Leedes Arabian | | |
| | | Mare | Spanker | D'Arcy Yellow Turk |
| | | | | Old Morocco Mare |
| | | | Old Morocco Mare | Morocco Barb |
| | | | | Old Bald Peg |

Old Bald Peg was by an Arabian out of a Barb Mare

The Pedigree of St. Simon, showing the colour distribution among the ancestors in the first three generations

| St. Simon (bay) 1881) | Galopin (bay) | Vedette (brown) | Voltigeur (brown) |
| --- | --- | --- | --- |
| | | | Mrs. Ridgeway (roan) |
| | | Flying Duchess (bay) | Flying Dutchman (bay) |
| | | | Merope (bay) |
| | St. Angela (bay) | King Tom (bay) | Harkaway (chestnut) |
| | | | Pocahontas (bay) |
| | | Adeline (bay) | Ion (bay) |
| | | | Little Fairy (bay) |

The Pedigree of Nearco, showing the four crosses of St. Simon in the first five generations

| | | | | |
|---|---|---|---|---|
| | | Polymelus | | |
| | Phalaris | | | |
| | | | Sainfoin | |
| | | Bromus | | |
| | | | | St. Simon |
| | | | Cheery | |
| Pharos | | | | |
| | | | St. Simon | |
| | | Chaucer | | |
| | Scapa Flow | | | |
| Nearco (1935) | | | | |
| | | | St. Simon | |
| | | Rabelais | | |
| | Havresac II | | | |
| | | | Ajax | |
| | | Hors Concours | | |
| | | | | St. Simon |
| | | | Simona | |
| Nogara | | | | |
| | Catnip | | | |

The Pedigree of Friar's Daughter, showing the three crosses of St. Simon in four generations

| | | | | |
|---|---|---|---|---|
| Friar's Daughter (1921) | Friar Marcus | Cicero | | |
| | | | | |
| | | Prim Nun | Persimmon | St. Simon |
| | | | | |
| | Garron Lass | Roseland | William the Third | St. Simon |
| | | | | |
| | | Concertina | St. Simon | |
| | | | | |

The Pedigree of Ribot. In each quarter there is a horse inbred to St. Simon, these being Havresac II, Apelle, Pharos and Papyrus

| | | | | |
|---|---|---|---|---|
| Ribot (1952) | Tenerani | Bellini | Cavaliere D'Arpino | Havresac II |
| | | | | Chuette |
| | | | Bella Minna | Bachelor's Double |
| | | | | Santa Minna |
| | | Tofanella | Apelle | Sardanapale |
| | | | | Angelina |
| | | | Try Try Again | Cylgad |
| | | | | Perseverance II |
| | Romanella | El Greco | Pharos | Phalaris |
| | | | | Scapa Flow |
| | | | Gay Gamp | Gay Crusader |
| | | | | Parasol |
| | | Barbara Burrini | Papyrus | Tracery |
| | | | | Miss Matty |
| | | | Bucolic | Buchan |
| | | | | Volcanic |

The Pedigrees of Havresac II, Apelle, Pharos and Papyrus, showing the inbreeding to St. Simon in each case

| | | | | |
|---|---|---|---|---|
| Havresac II (1915) | Rabelais | St. Simon | | |
| | Hors Concours | Ajax | | |
| | | Simona | St. Simon | |

| | | | | |
|---|---|---|---|---|
| Apelle (1923) | Sardanapale | Prestige | | |
| | | Gemma | Florizel II | St. Simon |
| | Angelina | St. Frusquin | St. Simon | |
| | | Seraphine | | |

| | | | | |
|---|---|---|---|---|
| | | Polymelus | | |
| | Phalaris | | | |
| | | | Sainfoin | |
| | | Bromus | | |
| | | | | St. Simon |
| | | | Cheery | |
| Pharos (1920) | | | | |
| | | | St. Simon | |
| | | Chaucer | | |
| | Scapa Flow | | | |
| | | Anchora | | |

| | | | | |
|---|---|---|---|---|
| | | | Sainfoin | |
| | | Rock Sand | | |
| | | | | St. Simon |
| | | | Roquebrune | |
| | Tracery | | | |
| Papyrus (1920) | | | | |
| | | Marcovil | | |
| | Miss Matty | | | |
| | | | St. Simon | |
| | | Simonath | | |

The Pedigree of Ribot contains five other crosses of St. Simon besides those found in the Pedigrees of Havresac II, Apelle, Pharos and Papyrus, making thirteen altogether.

The Pedigree of Gold Bridge, showing inbreeding to Orby

| | | | |
|---|---|---|---|
| Gold Bridge (1929) | Swynford or Golden Boss | The Boss | Orby |
| | | Golden Hen | |
| | Flying Diadem | Diadumenos | |
| | | Flying Bridge | Orby |

The descent of Petite Etoile from Americus Girl in the female line, showing the sire in each generation:

Americus Girl (by Americus)
|
Lady Josephine (by Sundridge)
|
Mumtaz Mahal (by The Tetrarch)
|
Mah Mahal (by Gainsborough)
|
Mah Iran (by Bahram)
|
Star of Iran (by Bois Roussel)
|
Petite Etoile (by Petition)

# *Bibliography*

This bibliography is not intended to be comprehensive; nor does it comprise more than a proportion of the sources which were consulted in the preparation of this book.

All the works listed have been consulted, and are suggested for further reading in the subject. Some may be read for enjoyment, others for instruction; a few may be read for both purposes, but it would be invidious to specify those which fall into this category. Fewer still may be read for the, perhaps perverse, pleasure of violent disagreement.


*The General Stud Book*. Volumes I–XXXVII.

*The Racing Calendar.*

*The Turf Register and Sportsman and Breeder's Stud-Book*, by William Pick, 1803.

*A Treatise on the Horse*, by William Osmer, Fifth Edition, 1830.

*The Horse*, William Youatt, 1831.

*History of the Turf*, by James Christie Whyte, 1840.

*The Horse Breeders Handbook*, by Joseph Osborne.

*Horse Breeding Recollections*, by G. Lehndorff, 1883.

*A History of the English Turf*, by Theodore Andrea Cook, 1901.

*Early Records of the Thoroughbred Horse*, by C. M. Prior, 1924.

*The History of the Racing Calendar and Stud-Book*, by C. M. Prior, 1926.

*The Royal Studs of the Sixteenth and Seventeen Centuries*, by C. M. Prior, 1935.

*Flat Racing*, The Lonsdale Library.

*Memories of Racing and Hunting*, by the Duke of Portland, Faber & Faber, 1935.

*The Breed of the Racehorse*, by Friedrich Becker, B.B.A. Ltd., 1935.

*Thoroughbred Racing Stock*, by Lady Wentworth, George Allen & Unwin, 1938.

*Animal Breeding*, by A. L. Hagedoorn, Crosby Lockwood, 1939.
*Farm Animals*, by John Hammond, Edward Arnold, 1940.
*Livestock Improvement*, by J. E. Nichols, Oliver & Boyd, 1944.
*The Breeding of Farm Animals*, by Chapman Pincher, Penguin Books, 1946.
*Breeding Thoroughbreds*, by John F. Wall, Charles Scribner's, 1946.
*Breeding the Racehorse*, by Federico Tesio, J. A. Allen, 1958.
*Bloodstock Breeding*, by Sir Charles Leicester, J. A. Allen, 1964.
*The History of the Derby Stakes*, by Roger Mortimer., Michael Joseph 1973.
*From Gladiateur to Persimmon*, by Sydenham Dixon, Grant Richards 1901
*Men and Horses I Have Known*, by the Hon George Lambton. J. A. Allen, 1963.
*The Bloodstock Breeders Review*, Volumes I–LXI.
*Stud and Stable*.
*The British Racehorse*
*Post and Paddock*, by 'The Druid', 1859.
*On The Laws and Practice of Horse Racing*, by the Honourable Captain Rous, R.N., 1850.
*The Mating of Thoroughbred Horses*, by H. E. Keylock, F.R.C.V.S., 1942.
*Family Tables of Racehorses*, compiled by Captain Kazimierz Bobinski, published by Lt-Colonel Stefan Zamoyski, M.B.E., 1953.
*Horse Racing*, by Dennis Craig, J. A. Allen, 1963.
*Horse Racing in France*, by Robert Black, Sampson Low, Marston, 1886.
*The History of Thoroughbred Racing in America*, by William Robertson, Prentice-Hall, 1964.
*The Roanoke Stud (1795–1833)*, by Fairfax Harrison, Old Dominion Press, 1930.
*The British Thoroughbred Horse*, by William Allison, International Horse Agency, 1907.
*Horse Breeding*, by B. Von Oettingen, Sampson Low, Marston 1909.
*The Brigadier*, by John Hislop, Secker and Warburg, 1973.
*The Statistical Record*, published by Weatherbys and Stud and Stable.
*Breeding To Race*, by Sir Rhys Llewellyn, J. A. Allen, 1964.

# Index

Compiled by P. Phipps

* Indicates there is an illustration of the horse